Delivering Safety Excellence

Delivering Safety Excellence

Engagement Culture at Every Level

Michael M. Williamsen, PhD
Cobden, IL, USA
cultureoddysey.com

Registered Office
John Wiley & Sons, Inc., 111 River Street, Hoboken, NJ 07030, USA

Editorial Office
111 River Street, Hoboken, NJ 07030, USA

For details of our global editorial offices, customer services, and more information about Wiley products visit us at www.wiley.com.

Wiley also publishes its books in a variety of electronic formats and by print-on-demand. Some content that appears in standard print versions of this book may not be available in other formats.

Library of Congress Cataloging-in-Publication Data applied for

ISBN: 9781119772132

Cover design by Wiley
Cover image: © Simone Hutsch/Unsplash

Set in 9.5/12.5pt STIXTwoText by SPi Global, Chennai, India

SKY10026147_040621

Contents

Acknowledgements

Katy Crane, an original owner of CoreMedia developed a friendship with Dr. Dan Petersen which led to my joining CoreMedia with her when they needed a replacement for then retiring Dr. Dan. Katy also started the drive to focus on safety culture and thus was one of the pioneers in this important field of safety performance excellence.

Tim Crane, my former partner and close and intense friend in CoreMedia. Because of Tim's leadership CoreMedia had the vision of beginning a safety culture consulting business unit. He also continued to fund the effort as well as push innovation and IT solutions for this new venture even in times of severe business uncertainty and cash flow struggles.

Brad Cosgrove was the CoreMedia, and later global manufacturing company, graphics technician. His innovation, skill, and perseverance kept us on the forward edge of graphics excellence.

David Crouch was one of my former managers at the global manufacturing company. David went through a similar effort to write his first book. His guidance and encouragement to my efforts were of great value to me.

Andy Schneider was the corporate global manager of safety for the entire worldwide company. Over the years Andy did an incredible job of working with the company's more than 500 worldwide manufacturing sites to reduce injuries. As the organization plateaued at a RIF of about 1, Andy searched for a company to buy that concentrated on safety culture excellence. Thank you, Andy, for bringing our small team on board and challenging and supporting our unique safety culture approach worldwide.

Aaron Janisko was a safety manger leader who has spent untold hours with me discussing and developing safety excellence under very difficult circumstances. Aaron is one of many such caring, engaged, intense safety professionals our team and I have had the pleasure of working with over the years. They have helped our process, and we in turn have helped their performance. It has been an incredible journey for all of us.

Dave Fennell is a recently retired Exxon Mobile Imperial Oil safety manager in Alberta Canada. Dave was one of the first safety professionals to embrace Dr. Dan Petersen's cultural approach to achieving safety excellence. Dave has been a good friend and a safety culture innovator for as long as I can remember. The risk tolerance material is all his.

Erik Williamsen, my son, did all the OCR (Optical Character Recognition) software conversion for Dr. Charles Bailey's Railroad Study booklet. The only surviving copy I could find of this embryonic safety culture work was a photocopy of the original book. It was barely legible, yet Erik spent hours doing the conversions of the text and excel graphics materials which now appear in the appendix of this book.

Mark Mays, a close friend from our days when our family lived in Colorado. Mark has deep interest and academic course work in creative writing along with an interest in writing a book. During a ski trip in Colorado, blizzard conditions snowed us in for three days giving us the time and venue to outline this book. Mark's ongoing critiques of my many drafts were of significant importance to my being able to finish this work.

John Busch, has had a broad professional career in teaching, research, and government service, including being the Chairman of Engineering at LeTourneau University in Longview, Texas. Dr. Busch edited my doctoral dissertation and thus insured its first pass acceptance with no changes by the dissertation committee of Columbia Southern University in Orange Coast, Alabama. John more recently spent many hours helping me with the final editing polishes necessary to get this book accepted by the publisher, John Wiley and Sons.

Raelee Williamsen, my wife of 50+ years and her brother Tom, who lives with us on our small farm, keep the television and radio on until 10 p.m. most nights. In turn, this provides the necessary incentive for me to move out to my man cave office in the barn and do the many hours of work necessary to complete this book.

Domino, our cat who walks around the farm with me in the dark after my day is done. I get solace from this furry friend.

Author Biography

Michael M. Williamsen has a BS in chemical engineering from the University of California, Berkeley; MBA from California State University Hayward, California; and PhD in business – dissertation topic: "How to Accomplish Organizational Turnarounds" – from Columbia Southern University, Orange Coast, Alabama. Certified Safety Professional (CSP). Since graduating from Cal, Dr. Williamsen has worked as a turnaround specialist with a number of small, medium, and large organizations throughout the world that were in trouble in one or more of their functionalities. Over the years the many experiences have helped him to originate and document a cadre of materials which have proven to be effective in developing a culture of interactive engaged employees and managers. The result is a vigorous culture in which the frontline people relentlessly solve their own problems by themselves with their own resources. In addition, these problem solvers have effectively taught and passed on the attitudes and skills learned within their companies and beyond.

List of Figures

Preface

The book teaches the use of in-depth, practical processes that enable frontline function people solve day-to-day weaknesses and dysfunctional culture problems in an organization. The author and his associates have tested and proven these culture improvement models and approaches in multiple industries worldwide. In the book's text each of the culture improvement processes and models are analyzed and worked through to conclusion. The solution is presented in dialogue format using case study snippets discussed between the author as the senior consultant and the organization's employees and leadership involved with the troubling issue that faces them. The case study snippets are from the author and his staff's actual experiences that have occurred in a spectrum of industries and organizations across the United States, Canada, Mexico, South America, Europe, Africa, Middle East, Australia, and India.

This book is different from others written about culture improvement. Every chapter has documented examples of challenging real world problems solved by actual frontline employees using simple effective tools that engage other employees in their group. These real examples are like the majority of workplace problems facing frontline employees on a regular day-to-day basis.

As you read this book you will encounter colloquial words and phrases with which you may not be familiar. There is a Glossary of definitions for these terms available to you by going online to this book's landing page in wiley.com.

Prologue

A while back a friend sent in the following thought-provoking question: "Do you see a reduced need for safety professionals in the future while considering the huge ongoing tech advances in the industry that could greatly affect the need and actions of present and future safety professionals? Examples include: robotics, drones, automation, employees using smart phones to capture hazards and send in reports, wearable devices that monitor a worker's health conditions and physical exertion, etc. With this tsunami of change, consider how technology is disrupting so many fields and causing job losses, e.g. trucking in the future with driverless trucks, delivery drones replacing drivers, robotic welding replacing welders, etc., how might this affect future safety professionals?"

One thing we can always count on in life is change. And with change comes transformation. Consider how the safety profession began in earnest with the Triangle Shirtwaist factory disaster back in 1911 (see Chapter 1). At that point in time there were no laws, or standards, or safety professionals. Then in the 1970s all the regulations and bureaucracies associated with OSHA caused another huge change in what safety was and how it is practiced. In the meantime, we have seen the rise and fall of behavior-based safety (BBS) and then the initiation of safety accountabilities and safety culture. Through it all we have transformed the way we work in order to protect others. And now our future seems to include many changes through the tech innovations mentioned and many more.

These changes bring to mind Moore's Law: Technically, Moore's law is the observation that the number of transistors in a dense integrated circuit doubles approximately every two years. The observation is named after Gordon Moore, the co-founder of Fairchild Semiconductor and Intel, whose 1965 paper described a doubling every year in the number of components per integrated circuit, and projected this rate of growth would continue for at least another decade. Since then, the term Moore's Law has been applied to other industries as an intentional generalization to describe the significant technology explosions occurring in many fields in addition to integrated circuits. These significant advances/changes will likewise require associated advances in the development and application of robust techniques for improving safety. Although we are no longer producing nearly as many (of what some would consider) obsolete technology products, there is always the need for the fundamentals of safety: such as regulations, PPE (Personal Protective Equipment), and the like including the continued development and application of safety fundamentals.

Newer technologies bring new challenges to other associated cultures such as human relations, training, industrial safety, and much more. The generational changes that come with a Moore's Law society also affect how we live and what we do as the older generations are continually replaced by younger generations. There are, and will be, foundational safety challenges that must continually be addressed. There will also be all kinds of new challenging safety issues with: electronic and chemical processing, nanotechnology, healthcare, robotic utilization, drone usage, biological safety, security enforcement, etc. The continuing tech upheaval does, and will, change what goes on in the field. As we look to the future, safety professionals, if they want to continue to protect employees, will need to adjust to the changes impacting our frontline production and society. A part of this future will be the need for safety documentation and accountability, and safety culture excellence which may very well be something challenging for younger generations to grasp. This newer generation has many people who have had far less practical experience than the older generations who grew up in a more "mechanical," hands on culture, which taught them the importance of personal safety, sometimes through the "school of hard knocks" and associated "ouch factors."

The need for safety professionals will still be an integral part of our world's future and a part of many ongoing transformations. Just as our profession has transformed from 1911 to now, it will continue to transform with technological advancements. Different skill sets, technical knowledge, and cultural approaches will continuously be required. There will be a new kind of safety professional required to meet the demands of an ever safer future with technology implementation. This is just like the truth that there have been huge changes in the skill sets, talents, and abilities of safety professionals now compared to when our profession was launched more than 100 years ago. Fortunately for this next generation of safety professionals, we all stand on the shoulders of those who have gone before us; we do not have to reinvent what they did. Looking back at the last 100+ years of our safety profession accomplishments this has always been the case. The safety profession will not go away, but we will have to significantly adjust and transform to the tsunami of technological change.[1]

How to go about this required transformation is the object of this book.

Much of the story line is anecdotal. However, it is also all based on real people and real happenings the author has experienced during his career of culture turnarounds for troubled organizations. The models contained in this book are meant to be thought provoking. There is underlying research behind them, but mostly they are the result of practical experiences of working with people who then begin to engage their own talents to identify the difficulties surrounding day-to-day life on and off the job and create needed improvements. There are some quotes from famous people including Albert Einstein, H.W. Heinrich, and Dr. Dan Petersen. However, they are not footnoted as this is a practical application book, not a text book. The author's objective has been to provide easily understandable visuals and context which will inspire hourly and salaried leaders to engage in and improve a culture that fixes problems and does not rely solely on fundamental initiatives which plateau way too soon (level 1 and level 2 tools explained later in this book, see Chapter 10). The author hopes this practical approach provides the inspiration, thought-provocative material, and tools for you to go beyond a reactive condition solution mindset and into personal accountabilities and responsibilities. This different kind of safety tool set will assist you and your fellow workers to apply the efforts necessary to achieve a zero incident/zero at risk activity safety culture mindset and its resultant performance. The reader is encouraged to dig deeper into the works of these and other respected people. Indeed, all the material in this book has been presented at numerous global conferences, and thus is documented on the Web. As is common to professions, this same material has then been modified (and sometimes improved) by others. As you see items of interest to you, e.g. training, look at the reference provided and then expand your search to get a look at the greater depth that exists and is constantly changing. The information is out there and in this book, all you have to do is let it in as it applies to your individual interests and needs.

May you both enjoy and benefit from this work.

Sincerely, "The Doc"

Note

1 Industrial Safety & Hygiene News, September, 2017, Vol 51, No. 9, p. 88.

Introduction

While I (the author, Mike Williamsen, PhD) was growing up my Papa was an hourly welder in the shipyards. At the end of each day he was bone tired and sometimes injured. I remember his wrist surgery, back surgery and a day when he went to an eye doctor who used a magnet to remove some weld slag from his eye. I never remember him complaining and yet his work-related difficulties made an impression on me. My mom and Papa lived through The Depression together. They never went to college, but both had the superb work ethic they needed to survive the many difficulties of their era. After I got a degree in chemical engineering from the University of California, I went to work in a petroleum refinery and then in an agricultural chemical facility. One of the important lessons I learned in the field of chemical engineering was the approach of focusing on *Unit Operations*. In both the classes and laboratories we focused on an individual unit operation, e.g. heating, pumping, distillation, etc., and then tried to optimize all the steps used in that process/unit operation. The unit operation analogy in safety could be something like how to be safe while working at heights, painting, lifting, handling hazardous chemicals, etc.

During my second job after graduation I discovered my interest lay in management rather than research or design. My wife agreed for me to go back to school and get an MBA thus better preparing me to go into the management ranks. My post MBA industrial life became one of turnaround work for the various organizations and industries who employed me. In one industry I was in charge of manufacturing engineering for a Fortune 20 company. In this role I was enjoying the endless challenges of working with plant and headquarters personnel as our small Continuous Improvement (CI/kaizen) teams significantly improved uptime and productivity for the 40 facilities I supported across the United States. In this role I made sure each small team used the unit operations approach of focusing on a single process such as: baking, frying, drive trains, logistics, and the like, and optimizing each step used in that particular process.

Suddenly one Tuesday my boss, Tom, told me about a fatality which had just occurred at one of the facilities. As two senior vice presidents were about to enter the plant, a lady violated a cardinal safety rule and entered an operating crane bay. While focused on her clean up tasks, the crane cycled and crushed this 38-year-old mother of three young children. The entire corporation was shocked. As they looked into their records they confirmed an even more shocking history of fatalities, dismemberments, and other serious injuries, which in the past had seemingly been accepted as injuries being an inevitable part of the manufacturing culture. Management prided themselves in being number one in their industry with respect to cost, quality, and customer service, and yet we were in the bottom third of our industry worldwide when it came to injury statistics. A decision had been made at the top that safety would become a measurable, compensable metric for management along with all the traditional measures of cost, quality, and customer service. That decision included that the safety metric must be brought up to world-class performance just like the others which were tracked. At issue was the fact that over the years upper management continually emphasized the company was a fun place to work. If this were so, the killing and maiming of employees must stop. After all, it was the frontline employees who produced the product which paid all of our salaries. With that crescendo, it had been decided I was to be in charge of safety for the corporation (of course in addition to my regular manufacturing engineering duties and at no additional pay). There were no restrictions on me or what it cost to accomplish this strategic goal of becoming world class in safety performance, though no one could define what world class was.

What a challenge: 10 000 manufacturing employees, 40 facilities strung across the United States, no safety staff anywhere, and only a reactive approach to the latest injury, no matter how serious it might be. I remember thinking "What am I doing in this role? I am an engineering manager, not a safety guy!" As I talked this over with my Papa one night, I distinctly got the vision I was embarking on a journey to save and improve the quality of lives of the likes of him and my mom. And I was all in!

The next steps included a series of jobs to practice all the background learnings I had absorbed, and then I met up with a small family owned company in Oregon, CoreMedia Training Solutions. And the safety journey to a "culture of correct" (developing an organization that lived a sustainable safety excellence commitment) shifted into high gear. As this safety culture engagement excellence process matured, another organization desired to use our tools in developing safety culture engagement excellence on a global scale. That far larger company purchased our very small family owned company and we became employees of a global Fortune 50 heavy industry manufacturer. This sequence of events brings to mind a recent YouTube video on the consequences of how our small, personal acts of kindness and engagement can have far-reaching effects on others whom we never knew our interactions affected. This message was presented as a metaphor about a person dropping a pebble into a pond and then watching the ceaseless ripples go out, with unknown impact into the unknown surroundings.

As I think about the number of people who have dropped pebbles into my pond, I am amazed how I was affected way beyond what was originally intended by the person dropping the pebble. Early on was a boss I worked for while attending graduate school. I was at a decision point to scrap a long planned graduation vacation with my wife, or go directly into the workforce and make money. I calculated all the financial ramifications and going to work looked very tempting. John, my boss at the time, then talked to me about a metaphorical high paying career of endlessly cracking eggs while sitting in a corner. He contrasted this high paying, mind-numbing job with seeking out what would deliver a lesser paying career in a field, which would bring personal satisfaction and not just more money. The vacation my wife and I took brought a personal experience and bonding that the extra money could never have delivered. The lesson in the trade off of more money versus a more satisfying personal life experience for the two of us and for our children has replayed (rippled) itself numerous times over the years.

Years later, Dr. Dan Petersen dropped his pebbles in my pond about the importance of culture and accountabilities in developing excellent safety performance for an organization. About the same time, other people in my life dropped some more pebbles in my pond related to creative problem-solving, continuous improvement, team excellence, and action item matrices. These ripples combined and resulted in the development of a safety culture excellence process that the Fortune 50 heavy industry manufacturer now uses worldwide, which, in turn, has helped to eliminate tens of thousands of serious injuries.

As a result of these people going out of their way to cause caring ripples in my life, I have had numerous opportunities to drop pebbles on how to deliver safety culture performance excellence with safety personnel and associated executives across our planet. Not surprisingly, the desire to help other people, as influential people have helped me, has provided many benefits for my many acquaintances. In turn, they have delivered on their personal desires to spread ripples of learning, way beyond mere safety-related issues, to many other people.

There are numerous others who have sent both pleasurable and painful ripples into my life. Getting to the point then, what pebbles can you drop into the huge pond of life, which will ripple out over time, to improve the lives of the masses of known, unknown and unseen others? That is the purpose of this book you are about to read (and I hope both enjoy and benefit from).

Part I

1

The Funeral

Aaron is physically sick to his stomach as he attends the funeral of a 37 year employee who fell to his death at work on the weekend. As he stands just behind the tearful widow, Aaron and his fellow employees are equally in tears. This was their close friend who was known for a good work ethic, reliability and friendship. Aaron, the organization's new safety manager, could see it coming with a Recordable Injury Frequency (RIF)[1] of >10 for more than a decade, and yet the company leadership just kept doing the same thing and hoping for different results. Aaron's day only gets worse as he feels the guilt of living in a sick culture of denial that has now taken the life of a good friend.

Do you ever experience something that is wrong, something that you try to hide? To some extent we all do! Personally, an experience such as this brings to mind recently working in a third-world country with a "challenged" work environment, while also traveling with family members after the work assignment. There were many excellent sights, people, sounds, and events wherever the vacation travels took us. And yet we experienced multiple troubles as well. While viewing a raging, dangerous river in a remote village the guide, Dalmiro, related that this was the location of a significant international extreme kayak event each year. Dalmiro then revealed that besides the boulders there was an added, hidden danger; the village of 10 000 or so people had no wastewater treatment and all the raw sewage was also a "secret" part of the raging river!

This "secret" comment brought to mind the story of a family member and her childhood obstinacy about eating certain foods. She hated hamburgers and refused to eat them. Her parents would "park" her at the table until she finished her meal. However, acting like the child she was, she crossed her arms and pouted. When her parents left the table, she would toss the meat behind the refrigerator and after a while call out to say she was done. All were happy as long as the subterfuge continued. One day her father cleaned behind the fridge, and the deception came to an end.

Delivering Safety Excellence: Engagement Culture at Every Level, First Edition. Michael M. Williamsen.
© 2021 John Wiley & Sons, Inc. Published 2021 by John Wiley & Sons, Inc.

Unfortunately many people in the safety profession have experienced organizations which have hidden the ugly, rotten, stinking truth about their culture of employee injuries. The subterfuge works for a while and then......

Give some thought to your personal and organizational circumstances. In the long run there is no escape from reality. You cannot hide the truth because untruths will eventually be revealed. Let us be ethical in all we do; you shall know the truth and the truth shall set you free. The upper management approach of Aaron's organization of hiding injuries was living in denial. Their solution to injuries was to send injured workers to Employee Relations (ER) for a multi-month review to see if punishment was warranted. This was truly counterproductive in many ways. Rather than focusing on what we all can do to eliminate a similar event from happening in the future, there were no reports of lessons learned, or issues resolved by searching out and identifying the actual blame. Additionally, the union and management both came to the same tragedy enabling conclusion – which was a lack of support for safety, and a lip service only approach to an understaffed safety department, eliminates trust and credibility. This denial approach only adds to the problem culture which continues to deliver the next series of painful injuries. Additionally, even if things do improve, beware, the lack of trust legacy hangs on for years. Our hourly and salaried people do not forget or forgive easily. Aaron has noticed that when there is an injury or mistake, there is always a contingent of the employees, at all levels, who immediately go back to the old paradigm of blame and shame. This included the ER function which was comfortable with the search for blame, and the potential for punishment. Change does not come easily.

The classic control, passive aggressive, old school challenges normally exist in these situations, and in other departments as well. Aaron's solution needs to not become angry, vindictive, or to go behind management's back. Rather, Aaron will have to persevere in upholding his values and his responsibility to do the right things that are effective in helping to resolve the safety and interpersonal issues. A part of this approach will require him to carry on a dialogue with the new incoming chief executive officer (CEO) and his staff. Aaron must use this method if he is to get them to support his desired approach to develop root cause solutions and a subsequent culture that includes a sustainable safety excellence commitment dedicated to significantly reducing injuries and associated incidents. It is no surprise that about 90% of these injuries happen in the operations group. As a result, Aaron will need to develop a solid adult-to-adult relationship with the operations hourly and salaried leadership personnel. Considering the history of the company, making such a turnaround in relationship excellence will not come easily. You will need slow and steady perseverance, Aaron.

After the funeral, Aaron is back at work and pulls out a report written by "the Doc," a consultant he hired to interview more than 100 hourly and salaried personnel in Aaron's organization of more than 1000 employees. The report refers to honest one-on-one input from the whole range of hourly through salaried employees who discussed their organization's safety and morale truths with the Doc. The employees did not rip and tear during the process, but they were brutally honest in their confidential comments. Aaron hurts as he reads and digests these painfully honest and ugly facts that he and others shared as inputs about their sick safety culture.

Aaron sits at his desk head in hand with disturbing thoughts going through his brain that: nothing is good, just another day/set of injuries to read and evaluate with no support for himself being the safety manager. Aaron is the leader of a small safety department which has a ½ administrative assistant time allocation, one safety resource up from the ranks, and two safety trainers, one of whom is on the ropes for his poor performance in other departments that got him transferred (hidden) to safety.

What kind of day lies ahead? Good = no injuries, or bad = one or more injuries. Aaron is up from the ranks. He knows the people requiring his injury investigations, and it mentally and physically pains him to do so. The company has been in business for more than 70 years and is one of the top 25 in the North American continent when measured by sales volume. For these same 25 entities they are 12th in size, but number 24 in injury rate with only an independent offshore business operation being worse.

As typical to industry, management gets paid on results for cost, customer service, and uptime. The company has had no fatalities or disabling injuries for quite a few years. As a result, the just retired CEO left a weak safety department and associated weak safety culture. They are complacent and multiple years behind what industry leaders are doing to prevent injuries. The safety Recordable Injury Frequency (RIF) has been greater than 10 for more than a decade. The former CEO's legacy approach for an injury was: a quick injury investigation; a secret report sent to Employee Relations (ER); followed by a secret and protracted/lengthy analysis as to what kind of punishment should be given to the injured employee as a result of any perceived negligence.

Aaron remembers a recent safety article that used the phrase Paradigm Paralysis. The focus of the article was a complaint about the tendency we all have of using old (and outdated) approaches to solve current problems. As Aaron reads the blog article he reminisces about a war hero acquaintance, Tom, talking about his career in the armed forces. Tom's observation referenced military leadership's oft-used approach of employing the same tactics for the next war that they used in the last war. Tom's conclusion was that this approach just does not lead to optimum performance, in war – or in safety.

Our safety profession history began in 1911 with a disastrous, multiple life-ending tragedy at a New York garment manufacturing sweat shop (*Triangle Shirtwaist Factory fire*). Over the ensuing years "we" have experienced all kinds of research, regulations, techniques, technologies, leadership, education, training, and the like. Much of this information (but not all) has moved us to better downstream indicator safety performance.

Talking with past generation safety people, there is often a great reluctance to try new safety concepts that are outside of their experience comfort zones, ergo, Paradigm Paralysis. Certainly, the foundational approaches which have been developed in the past 100 years still apply. And yet, this decade's safety performance plateau is not satisfactory. We must relentlessly pursue better techniques and tools to eliminate the possibility/probability of injuries/incidents.

Our current war on injuries and incidents is being fought by a new generation with new cultures, different workplaces, and a myriad of other differences from what the older generations experienced. We must be open to considering and trying new approaches which can help us win the important safety battles that face us now and in the future. And yet government and some industry safety bureaucracies seem to often stick to the use of regulations followed by punishment as the predominate model with respect to safety improvement. In truth, a very conservative approach is influenced/hindered by the "standard practice" approach that is greatly influenced (hamstrung) by the litigious nature of society, i.e. not trying something out of the ordinary in order to minimize lawsuits! Such, "Standard Practice" cultures built on conservative tradition can be VERY difficult to change.

Since the 1970s' Occupational Safety and Health Act (OSHA) became law, OSHA has tried a number of approaches in an effort to improve safety in the United States:

- The regulations have set a foundational standard that has definite merit.
- The punishment by legal fines structure got some corporate attention, but it has led to a negotiating game which does not have its focus on improving safety, merely negotiating cost.
- Unannounced on-site inspections have had little to no discernible impact on personnel safety rates. It appears that OSHA inspectors, with little in-depth knowledge of a company's real hazards, lack credibility and instead often deliver derision.
- The Voluntary Protection Program (VPP)[2] system had merit, as it focused on assisting those who were seemingly serious, to improve their regulations compliance.
- The shame and blame approach only seems to anger the guilty, while adding glee to the segment that revels in a seeming punishment to corporate entities.

Untold billions of dollars spent on OSHA have resulted in minimal improvement in personnel safety numbers. The plateau in safety performance is not improving with a "trouble equals government/business leadership punishment" model. A number of safety professionals and managers committed to safety excellence, who have experiences in various industries in multiple countries and cultures, have settled on a better working model. This approach is more along the lines of a safety culture where "trouble equals value added assistance." Subsequently, if the leadership cannot improve performance when given such assistance, their poor performance leads to a change in leadership.

Details from such innovative accountability-based safety cultures are revealed in a significant number of large global companies. These organizations have done far better in safety performance by definitely employing manufacturing fundamentals while also improving their safety culture. They have discovered the need to go beyond the "one trick pony regs (regulations) and punishment models." An easily available search approach would reveal the industries, cultures, and locals which need focused assistance. They are likely the same ones that traditional approach only leaders think are in need of more of some kind of punishment. "High injury rate plateau organizations indicate the beatings will stop when the safety performance improves" model, is not effective in the long run.

OSHA birthed the value-added regulations fundamentals by copying (and adjusting) the policies, processes, and procedures of companies which were successful in safety. The models that successful companies have used in improving their day-to-day safety performance work for the laggards as well. Across the board, engaged safety leadership which goes beyond the necessary strong regulations base drives a safety culture of excellence. It is time to try a similar approach for improving safety cultures by copying and adjusting what has been shown to work and applying this model to those company cultures that are in need of value-added assistance.

Over time I have been faced with a few new OSHA directors who view themselves as "The new sheriff in town." A common thread of the new sheriff is a promise to punish industries as their way to safety success. At the end of their term I have had difficulty seeing any real statistical difference in safety performance. Since the 1970s there have been incredible continuous improvements in multiple technologies worldwide which have delivered amazing performance improvements in just about all that we experience and do. That is except in safety, where the old, tired, low performing approach of the seventies remains the outdated norm. Continuing the beatings, which have proven to be unsuccessful in improving safety performance, is a kind of leadership (either governmental or industrial) insanity which needs to be changed. Doing the same thing and expecting different results just does not make sense.

As Aaron reflects on the above safety culture reality of his company, he has also reached the conclusion that the union which represents the workers is a part of the poor safety performance culture where he works. They are safety complacent too because of a good technology apprentice program, no critical injuries for a number of years (except for now), and a good company medical benefits system which pays for fixing the injured employees and gets them back to the job. However, in Aaron's company there is a huge separation/gap between frontline employees and upper management. The high injury rate, a lack of caring, along with a focus on punishment with no real action by management, has become an angry boil that keeps the safety department as the object of festering unhappiness by all.

The general feeling across the organization is that: this is a great place to live; the pay is good; the work is not all that hard; and the frontline employees are well trained, highly skilled and have a good work ethic. However, the huge gap and lack of respect between the frontline employees and management have morale in the tank and skilled frontline employees leaving for other companies who need their skills and have a better culture.

Aaron realized doing the same thing and expecting a different result just does not make sense. But he is up from the ranks with little safety background and his upper management just does not seem to care or want to get involved in anything that is troubling. There is no apparent silver lining in the dark foreboding clouds surrounding the safety reality. He reflects on the safety filter[3] discussion he had with the Doc.

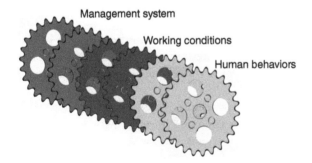

Figure 1.1 Work on the job site. Source: Reason, James (1990). The contribution of latent human failures to the breakdown of complex systems. *Philosophical Transactions of the Royal Society of London. Series B, Biological Sciences.* 327(1241): 475–484.

Safety systems have barriers (filters) which help prevent injuries and incidents. These include management systems, working conditions, and human behaviors. In a 24/7 operation, or any other for that matter, these filter plates are spinning as the work dynamic and process relentlessly continue. However, each filter piece has some holes (weaknesses) in it, and when the spinning holes line up, an incident occurs. When someone is in the wrong place, at this wrong time of weakness alignment, an incident (near miss/close call) or an injury (hit) also occurs. Aaron realizes that his employing organization has way too many holes in its safety barrier filters.

He remembers his friendship with an excellent skier, Tim, who gave him a great lesson in the facts of achieving performance greatness. Tim loves skiing moguls and by his own standards is "pretty darn good at it." In an effort to improve his abilities, he paid for coaching from a professional skier. Much to Tim's chagrin the pro's evaluation of his "pretty darn good" technique and ability came back as "a closely linked series of recoveries." They had a good laugh, and then discussed another acquaintance who decided to become a pro golfer even though he never really played golf until later in life. His professional coach advised a need for 10 000 hours of concentrated practice to raise his skills to a point where he could make a valid decision, whether or not "to continue to try and become a pro golfer."

Tim and Aaron reached some conclusions out of their dialogues about professional performance. You need dedication, practice, drive, ability, good technique, and a relentless pursuit of excellence to even come close to the execution levels of professionals. These dedicated professionals are able to daily achieve measures accomplished by only the best of the best. Or in street language; "My commitment is stronger than a bumper sticker, but less than a tattoo" is just not good enough to get anywhere near great results.

Is there any parallel in safety performance? You bet! Those organizations which routinely go years without lost time or medical incidents have a leading edge engagement culture that has their entire organization focused on dedication, practice, drive, technique, and daily safety leadership development at all levels. This kind of safety culture delivers an end result which visibly demonstrates a relentless pursuit of zero errors (incidents). Every day they practice and live models of process excellence in operations AND safety. They are always in search of ways to improve their performance in every aspect of what they do, **including safety**. They do the fundamentals well and then go way beyond the basics. The rest of the pack of safety professional wannabes, who have safety cultures that are pretty darn good, seem to just live a culture where there is a closely linked series of recoveries instead of emphasizing a culture of prevention excellence.

Aaron racks his brain. It all seems so hopeless. He begins to dig deeper on many of the other road blocks that stand in his way of stopping the injuries of his friends in the field. Aaron needs to make a choice: "Never, never, never give up" thank you very much Winston Churchill or "When you wake up tomorrow you will still be ugly."

Notes

1 **RIF** *Recordable Injury Frequency ([number of injuries requiring medical/doctor treatment] x 200 000) divided by total hours worked.*
2 The Voluntary Protection Programs (VPP) recognize employers and workers in the private industry and federal agencies who have implemented effective safety and health management systems and maintain injury and illness rates below national Bureau of Labor Statistics averages for their respective industries.
3 James Reason (2000) proposed the image of "Swiss cheese" to explain the occurrence of system failures, such as medical mishaps. According to this metaphor, in a complex system, hazards are prevented from causing human losses by a series of barriers. Each barrier has unintended weaknesses, or holes – hence the similarity with Swiss cheese. These weaknesses are inconsistent, i.e. the holes open and close at random. When by chance all holes are aligned, the hazard reaches a person and causes harm.
4 This additional publication material is "Reason, James (1990). The contribution of latent human failures to the breakdown of complex systems. Philosophical Transactions of the Royal Society of London. Series B, Biological Sciences. 327(1241): 475–484".

2

No Support for Safety

> *The week after the funeral Aaron tries to de-stress by going out to dinner with his wife. As he sits down the images of the funeral, the fellow workers, the widow… flood into his mind. As he crumples into the chair, internal emotion erupts and he bangs his head repeatedly on the table. Finally, he gets back into some small level of mental control, but couldn't eat. They apologize to the waiter, and Aaron's wife drives toward home as he struggles with the impossible task staring him in the face: neither management nor labor is willing to do anything to improve their miserable safety record. Sure, there is no lack of talking heads and condemnation on both sides. However, a decade of complaining without any substantial actions has delivered absolutely nothing (no improvement) to any real safety culture or employee injury reduction.*

Back at work, Aaron chairs this month's joint safety committee meeting. As ever, the union and management safety leadership sit on opposite sides of the table and the joint safety meeting becomes a classic grievance meeting. The union vice president pulls out a three-page typewritten list of complaints that has not changed substantially from last month's list, or the many similar ones from months before. Two hours later the bickering comes to an end as the lunch hour signals the end of another worthless union-management monthly joint safety battle. Aaron looks over the latest union three-page condemnation of management inaction. He cannot blame the union for continually bringing up the lack of any action on numerous small and large types of items such as a lack of adequate lighting, or no training of emergency crews who participate in emergencies, or, or, ad infinitum. The union leadership rightfully believes this lackadaisical, no action management culture to be the precursor to injuries and near misses/close calls which continue to occur. Aaron mentally reviews the OSHA 300 log and sees numerous repetitive injuries which could have been prevented if they would have spent some discretionary dollars on simple things such as lighting, barricades, and correct boom tool attachment.

Delivering Safety Excellence: Engagement Culture at Every Level, First Edition. Michael M. Williamsen.
© 2021 John Wiley & Sons, Inc. Published 2021 by John Wiley & Sons, Inc.

In truth, management wages a constant battle for any safety spending to support this 1000+ employee company. In their view safety is a cost center which lacks any Return On Investment (ROI). Aaron replays the management diatribe in his mind: "We just pour dollars in while counting injuries, Workers Compensation (WC) costs, and payroll costs." He knows that WC costs have averaged up to and beyond a million dollars/year for more than a decade. Why cannot management understand that fixing these small and large problems would save an incredible amount of money and improve net profit? These fixes would also eliminate the intense amount of pain experienced by frontline workforce employees as a result of last year's 39 lost time injuries. Additionally, these fixes that result in many fewer injuries would also greatly improve employee morale. Why cannot the union leadership pitch in and fix some of the issues that face them every day, instead of making up next month's "no-action list" and rubbing it in the face of their opponents? Aaron begins to realize a sports analogy: there is an intense offense and an intense defense, but no team work. Consequently there are no victories to celebrate because status quo is all that is expected. The organization continually lives a WYSIWYG culture (what you see is what you get) as the two dysfunctional organizations routinely lock horns in an ongoing rutting contest, which just keeps rolling on at a miserable Recordable Injury Frequency (RIF) of 10+.

Aaron decides he must do something different to get out of the culture of insanity, but what is it? They follow the OSHA rules and guidelines, yet the painful RIF plateau of 10+ never gets better. Out of frustration Aaron reaches in and pulls out the Doc's business card which has collected dust for a number of months in the dark inner reaches of his desk. He hears himself mumble "But what use is this? I don't have any dollars I can spend on this kind of help." "Never give up, or ugly tomorrow? What have I got to lose?" as he dials the Doc's number.

It is a long and intense phone conversation between Aaron and Doc. As it ends, Aaron comes away with some personal revelations and a simple model that should help him begin the teamwork necessary to deliver real solutions to the monthly inaction list. Aaron looks at the notes he took during the call. He scratches out "notes" and gives the list a new title, "Revelations":

- **People do not care how much you know until they know how much you care.** That said, management does not need to be the smartest people in the room. We need to use everyone's abilities if we are going to have any chance of being a team.
- **Culture trumps process.** It costs nothing to show genuine appreciation for the good things people do every day. When a person is asked to look into an issue, their actions need to be followed up with genuine appreciation and reinforcement of their doing the right actions which move toward a solution. This kind of positive feedback helps reinforce a new culture of action; one of complaint

equals goal, which is far different from the current culture of complaint equals BMW (Bellyache Moan and Whine).

- **Act on employee concerns.** Acting on the many small issues which crop up over a month does not really cost enough to affect the monthly bottom line accounting numbers. However, this kind of frequent action of fixing what is bothering your people does genuinely affect your employees' attitudes and behaviors. There is a simple tool, an Action Item Matrix (AIM, Chapter 13), which easily and effectively helps people track what needs to be worked on and what has been accomplished. The first AIM needs to go to the maintenance team union employees with a monthly bogie of required hours to be spent on safety items. Look at the safety back log and pick a number of hours we can begin with: more than 10 and less than 100 is a good guideline for the launch of this initiative.

1	Outcome	ACTION ITEM/TASK	WHO	TARGET DATE	COMMENTS	COMPLETED
2						
3						
4						
5						
6						
7						
8						
9						
10						

(a)

Outcome	Action item / Task	WHO	Target date	Comments	Completed
1. Lighting	Outside area of maintenance office is dark	Fred	March 15	Talk to electricians, get cost and time estimate	
2. Bucket truck safety	Tooling attachment is wrong	Erik	May 14	Contact vendor and other of their customers with the same problem	
3. AIM list development	Develop an honest punch list	Aaron	Next week	Interview each work group	
4. Management involvement	Work with the new CEO	Aaron and Doc	Next week	Doc to set up meeting	

(b)

Figure 2.1 Action item matrix.

- **Trust and credibility are mission critical; there are no secrets in any organization.**
  - ○ Your employees at all levels know what the truth is and that affects their performance
  - ○ When your employees feel empowered, they do not want to let you down
  - ○ When you do what is right, your employees will do what they can to do what is right in return for your efforts
  - ○ If you do not fix what is needed, they will provide no help in return for your lack of efforts, and the organization becomes one of RIP (Retired In Place). This kind of culture acts like one that does not really care, and that kind of attitude is the death knell to any kind of improvement possibility

> *Aaron looks at the list and the Action Item Matrix* (Figure 2.1a and b) *the Doc emailed during the conversation. After a few minutes of additional conversation, he mentally begins to fill in action items he knows they can complete. And then he stops, prints some copies of the draft AIM and goes out to talk to the employees he knows who are willing to participate in the required teamwork. Without exception, those who are willing to engage give him items which are sticking in their craw. Some are difficult, like changing the tooling on the bucket truck booms and some are so simple, like additional lighting. He remembers another comment the Doc made; "The minimum level of expectation becomes the maximum level of performance." Aaron realizes all that was needed was his proactive leadership. Aaron and the Doc started the AIM* (Figure 2.1a and b). *There will be many more items added as the "Find it, Fix it, Move on" initiative progresses. As Aaron drives home he is looking forward to taking his wife out to a dinner. While in the background he considers the beginning of a plan to improve the safety culture and stop the inexcusable injuries.*

3

The Tyranny of the Urgent

Aaron sits down in his office after a week of talking to employees at all levels of the organization about what needs to be done to fix their safety culture. It is no surprise that his action item list includes about 90% condition fixes. After all, the regulations focus on unsafe conditions. There are some cultural issues dealing with the huge disconnects between management and labor, and even a few suggestions as to where and how to begin the needed fixing and healing work. But, the 150+ condition shortfalls are truly overwhelming as to where to begin considering the limited resources available. This approach seems to be what the people want, and hardware issues seem to be a logical starting point, but how to prioritize the many items in order to override the tyranny of the urgent and provide direction for Aaron's small team to begin its needed work. He next goes to his own tried and true soul searching method; puts his head in his hands on the desk and thinks for a few minutes. He then straightens up, flexes his muscles and calls the Doc.

They schedule another two-hour time block that, when it occurs, kicks off another practical teaching. This one-on-one event is all about how to improve a dysfunctional culture which has too much to do, for too few people, with too little time available for fixing all the needs.

The Doc shared his many years of experience working as someone who came into a company or organization which was in serious risk of going under (bankrupt). It seemed strange to the Doc that, almost without exception, there was a significant resistance to making change/improvement in a culture that most people thought to be terminally sick. Since he was the change agent, he experienced lots of resistance from these organizations which had become comfortable with their weak status quo. After a few rounds, the Doc came to a personal conclusion that most of these struggling cultures fit a 5-5-90 model:

Delivering Safety Excellence: Engagement Culture at Every Level, First Edition. Michael M. Williamsen.

- Five percent of the people would do whatever they could to help you make the necessary improvements, the "developers."
- Five percent would do whatever they could to fight any changes, the "resisters" (or CAVEmen [Citizens Against Virtually Everything]).
- Ninety percent would go in and out with the tide as organizational ebbs and flows occurred. This 90% group had a common motto: "Just tell me what to do so I can do my job." They were a picture of sheep following whichever shepherd was in charge.

In the beginning of such a turnaround assignment the vocal, aggressive, negative group always had the upper hand in a weak organization with poor leadership. Thus, the turnaround leader seemed to always engage in an uphill battle, kind of like the legendary Greek mythology character, Sisyphus, who was condemned to an eternity of rolling a boulder uphill, then watching it roll back down again. Only in our safety culture scenario, there are 95% of the organization's people pushing back against the needed uphill battle against change.

The vocal and aggressive "resisters" seemed to never let up. The Doc shared about the time he told his wife that he was like an elephant on a tightrope performing in a circus. The bad news was that there were people in the stands paid to snipe at him and that being a big target he took hits on a regular basis. The good news was that he had a thick skin and would keep relentlessly moving forward, provided he could keep his focus on the goals of what needed to be done.

There was a lesson from this that has stayed with the Doc. At first the Doc frequently got angry with the Cavemen who were always a thorn in his side. Then one day at a lunch a good friend calmly listened to the Doc's litany of complaints and associated anger. As the Doc paused briefly "to come up for air" the friend shared a life-changing lesson: *weak people avenge; strong people forgive; intelligent people forget*. Additionally, his friend shared another of his observations, these trouble makers would continue spreading their venom and one day their attitude and poison actions would return to cause them significant trouble. "What goes around, comes around." "A man cannot escape his past. The best he can hope for is to outrun it for a while." The Doc then added "Aaron you should keep these lessons in mind. As a change agent you will always have back biters, or worse, to deal with." "Put your energy where it will benefit yours and the organization's needs. Anger only steals your ability to focus and make the necessary improvements."

In order to have a rapid turnaround success the Doc discovered that his and the organization's efforts and scarce resources needed to be focused on working with the "developers." These are the people who are able to participate in actively digging themselves out of the poor performance holes. As the efforts to improvement begin to take place there is always too much to do, by too few people to do it. And thus the telling question Aaron called about was the constant question. "How

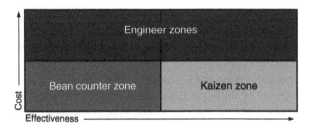

Figure 3.1 The ROI matrix.

to effectively and efficiently prioritize the mountain of 'to do items?' The Doc's answer was that through the years a simple 2×2 matrix, termed an ROI matrix (Figure 3.1), provided the necessary focus. The ROI uses no upper division math and has worked in every circumstance where this priority difficulty has existed.

ROI – What to focus on in the world of never ending things to do.

Over the years the Doc has dealt with a large number of organizations that have needed huge amounts of improvement. When their people honestly wrote down all they needed to improve, the list could fill several flip charts/white boards. Despair would then set in as the hourly and salaried improvement team leaders recognized there was too much to do and not enough time, money, or personnel to do all the work that was glaring at them in the face. Once these groups began using the concept of a simplified *ROI Matrix* they were easily able to prioritize the numerous difficulties, without using the typically challenging economics math.

To use this concept, the facilitator makes a 2×2 chart like the one in Figure 3.1 which shows cost versus effectiveness. After some lighthearted give and take, they are able to determine what high cost and high effectiveness looks like to the front-line dominated organization. The facility manager gets put on the spot and has to answer the question of how much funding they can locally provide and what kind of return they would expect for that amount of money. An example would be something like "$10,000 and the elimination of at least one medical injury." The team now has the management's approved upper limits set for the *ROI Matrix* team exercise.

The upper right-hand quadrant (Figure 3.1) is for the high-cost, highly effective solutions. This usually involves a fair amount of capital expense, like purchasing and installing new equipment. The upper left-hand quadrant is the high-cost, low-effectiveness zone, in other words, "I thought I was doing the right job by spending this money, but indeed, I was not." The Doc called these two the "engineer zones" because generally high-cost items are more technically challenging in nature and engineers often like to work on technically challenging projects. The lower left zone is the low-cost, low-effectiveness solution. He affectionately called that the "bean counter" or "accountant zone." A common paradigm deals with money people seemingly being interested in cheap, but not necessarily effective solutions. The lower right low-cost, highly effective quadrant is the "endless kaizen zone" ("kaizen" is a Japanese word that translates to small changes forever, or continuous improvement).

As the team focuses on the many potential solutions to its problems they put up an *ROI Matrix*. Each member of the team talks about a particular solution which interests them. The group then engages in an energetic discussion, as to both cost and effectiveness. Examples could include things like painting guard rails, launching an observation program, fixing a weak incident investigation system, etc. The team then cooperatively works together to place each one of the potential solutions in its appropriate location on the *ROI Matrix*. As this teamwork exercise progresses, it quickly becomes evident where to focus the scarce available resources. Thus the team uses a very effective visual mechanism which drives overall engagement and agreement for quickly sorting and prioritizing work to be carried out by the continuous improvement teams.

Aaron receives his challenge from the Doc: "If you have too much to work on, with too little time and financial resources to do so, why not try this kind of *ROI Matrix* approach?" "And who do you work with using this approach?" "Most of your time needs to be spent leading the improvement efforts with those who would engage with you to accomplish needed tasks, and you just have to put up with the incoming negative efforts of the 'Cavemen'." "Welcome to the real world of organizational change Aaron!"

The good news is that after about nine months of using this approach, improvements begin to take hold, performance improves, and the culture begins to make a notable shift. Quietly and methodically over time the "Cavemen" no longer have the upper hand. The 90% follower group begins pushing back when told to resist change, and they slowly become advocates of improving the organization and its efforts. The needed sea change will then be visibly and palpably upon you when we (both hourly and salaried leadership) engage our people in focusing on and fixing what they know to be wrong, while we let the flak throwers fade into the background. It is like driving a vehicle which requires concentration and attention on the road ahead (the goal) because no progress is made while concentrating only on a rearview mirror!

While all this build up work to improvement is going on, the 800 pound production/operations gorilla always takes precedence. There are numerous examples of dangerous shortcuts which are taken in the field to get production/customers back on line. Aaron thinks back over the last year's injuries and ticks off more than a dozen of these aggravating, painful examples. All are indicators of the weak safety culture that exists at every level of the organization. None of the parties in his organization are without blame, all share a piece of the blame, including Aaron's safety department.

Yet there must always be time made available for the ineffective, boring, regulatory required safety training done by boring, poor performing safety trainers. And when things go wrong because there was not time to do the job right, there always seems to be time to redo what should have been done correctly the first time. These frustrations will continue to occur, but at a lesser frequency after nine and more months of fixing stuff people have been accustomed to just living with. And of course there is more bad news; while in the embryonic state of organizational culture improvement, injuries will stop production/operations. In turn, these happenings will always put the safety department in the role of the safety cop bad guy.

The ROI matrix is a useful tool. However, it takes serious leadership to engage in this kind of journey that lights a fire for improvement. The Doc closed this time together with a challenging thought/concept; "Management is lighting a fire under someone. Leadership is lighting a fire in someone." "The ROI matrix approach is an important first step for an organization to develop the leadership that can replace a culture of ineffective management."

> *Aaron hangs up the phone and goes for a walkabout at a few of the company's field locations while he cogitates about all the input teachings he has just received. That evening he strikes a small smile as he quietly tells himself; "It's time to suck it up buttercup!" And thus, the next day Aaron decides to act, and then he begins building his case to implement a change in his company's culture. Therefore, he begins sketching out a mental plan to develop interactive, original training which satisfies the legal requirements and also engages the hourly employees. In the background he quietly tells himself: "Great idea, but how am I going to do this with no significant safety budget, no significant hourly or salaried leadership that is supportive, and a cadre of strong resisters/cavemen who roadblock my attempts for safety improvement?" "I guess it's time to break out the Action Item Matrix (Figure 2.1a and b) and bring my team to an ROI matrix (Figure 3.1) brainstorming event."*

4

No Pay for Safety

Aaron pauses before diving into the long "to do list." There is still one immense obstacle to overcome: a total lack of management participation. He remembers the Doc telling him about the few Critical Success Factors (CSF) which determine management pay, and therefore drive its active participation. At Aaron's organization the big three management Critical Success Factors are all wrapped up in the operations indicators of cost, quality and customer service.

Aaron recognizes that before he gets totally engaged with the AIM and ROI matrix tasks he needs to have a parallel tactic which will engage upper management in improving the safety culture. But how can he affect management pay when it comes to safety? It's time for another two hour phone discussion with the Doc.

The Doc reminisces about his being the new "safety guy" working with the famous and very experienced safety consultant, Dr. Dan Petersen (see appendix B for details). Immediately upon hiring Dr. Dan as the safety consultant, the Doc was exposed to an incredible amount of new material engineers were not taught in a university curriculum. After 42 years in safety Dr. Dan Petersen, who is often referred to in trade publications as a "safety guru," concluded that safety incidents were due primarily to management and management system failures. In his mind OSHA was not at all the culprit that upper management often thought them to be. OSHA was doing their job of providing a focus on the conditions that have often led to serious injuries and fatalities. However, management was another issue altogether and thus to quote an old-time cartoon character Pogo Possum, "We have met the enemy and they are us."[1] Dr. Petersen explained his safety philosophy as being centered on leadership and accountability as the way to develop an organization's culture of safety. In a 1996 interview with EHS (Environmental Health and Safety) Today magazine, Petersen said that managers must learn that they have

Delivering Safety Excellence: Engagement Culture at Every Level, First Edition. Michael M. Williamsen.
© 2021 John Wiley & Sons, Inc. Published 2021 by John Wiley & Sons, Inc.

to do things on a regular basis to develop the culture which produces safe behaviors. He went on to profess that he did not know what those things were, but that organizations must unleash their management skills on safety problems.

One of Dr. Dan's many paradigms was: "What gets measured is what gets done and what gets rewarded is what gets done first." Back then, when the Doc was newly in charge of safety for a Fortune 20 company, upper management was paid significant bonus dollars for excellent performance in cost, quality, and customer service. There was no incentive for safety performance and consequently very little attention paid to personnel safety numbers. Neither the Doc nor Dr. Petersen were proponents of safety bonus payments based on injury rates, the lagging indicators which are driven by what you do not want to occur. However, back at that time leading indicators were almost unheard of, so the safety culture improvement group did develop a simple bonus structure which gave significant emphasis on reducing injuries at the plant manufacturing level. In the past, 1/3 of the bonus pool was being paid out for excellent performance in each of cost, quality, and customer service objectives. The new plan, approved by an upper management which wanted to end their miserable safety culture, was 25% of the bonus pool for excellent performance in each of cost, quality, customer service, and *safety culture and injury rate* objectives. The safety incentive system was based on:

- Practical, effective safety accountabilities for all employee levels of the operations groups in manufacturing facilities.
- The implementation of a continuous improvement culture which engaged hourly and salaried employees in fixing whatever was not believed to be correct, or was complained about, with respect to safety.
- A safety work order system which dedicated the necessary resources to fix condition issues.
- An audit system which was not based on the typical OSHA paperwork approach. Rather, there were open-ended interviews with hourly and salaried employees which got to the core of safety accountabilities being regularly and credibly performed by all levels of personnel at the facility. The audits also included some records of safety continuous improvement/kaizen teams and safety work order attention and closure.

There was no credibility among the field facility managers that this was "the new normal." After all, they had never even been evaluated on safety for all the years the company had been in existence. That was until the first quarter payouts occurred! Suddenly there were 40 plant managers wanting to know what to do and how to do it "because money talks." Even with this significant incentive only four of the managers agreed to put forth the needed efforts to try the new approach in year one. The unit operation of safety culture excellence was a work in progress and they became the necessary developers (guinea pigs) on which to try "all this new safety culture stuff." At the end of the year, their performance improvement was spectacular. In year two of the new safety culture initiative, seven more plant managers signed up for this accountability, continuous improvement, engagement type of safety culture, and they too did extremely well. However, the remaining plant managers seemed unwilling to take on a safety goal and yet still got ¾ of a good incentive payout, which seemed enough for them without having to go through the hassle of "another corporate program." Once again, the safety culture improvement team received excellent backing from upper management as senior management issued a decree that went along the lines of "Safety excellence is a condition of employment for you to continue to be an employee of our company." But it took a fatality and three years of extreme effort on many fronts to get to the managers to all "buy in to" (accept) the program. Clearly, safety incentives were, and still are, a major issue with which organizations need to contend.

In the classic operations, organization safety, if it is a part of the pay plan, it is small or minuscule when compared to the big three. If safety is a part of pay for performance, it is typically focused on the reactive negative effects of injuries, like RIF and workers' compensation payouts. Seldom does the classical safety pay-out have anything to do with requiring proactive accountabilities by management, frontline employees, and supervision. Mostly, classical safety seems to be viewed as being lucky or unlucky.

This pay roadblock often seems insurmountable. Upper management says it cares about not injuring the employees, but there is no visual or activity based-evidence, only lip service. The union continues to vocally attack management and demand management personnel and/or attitude changes, but to no avail. How then can Aaron and his team overcome this significant safety improvement barrier?

Aaron considers his options and grows despondent over the impossibility of effective implementation. Why not just be like the rest of the organization and join the complaining group? The Doc warns Aaron that this approach is to no avail other than alienating upper management. In fact, joining the griping culture will hurt his efforts with:

- *Upper management, since Aaron will become just another of the whiners who helps destroy morale and performance, but does not engage to an extent in what helps improve performance*
- *Supervision, since they have no need, desire, or respect for anyone whose modus operandi is BMW (Bellyache Moan and Whine)*
- *Hourly employees, since those who were trying to help in improving the situation and culture are already overwhelmed by the resisters who for 10+ years have had the upper hand*

There is just no value being added to resolving the noticeable shortfalls by joining a group that desires to tear down suggested changes without delivering any of their own appropriate ideas or plans for implementing the many needed improvements. *The Doc warns that joining the BMW group would likely be a fatal career decision for Aaron. The Doc mentions that in his past job situations he has made some poor decisions which came back to bite him. Each time he did so, the result was the Doc got to "eat his own plate of dirt." He tried to make sure he learned these lessons so he didn't have to eat the same plate of dirt twice. Aaron shudders at the plate of dirt analogy yet is thankful for the good counsel.*

There has been a very recent CEO change which brought in a new leader, Craig. In last week's staff meeting Craig outwardly publicized his belief that personnel safety is the number one item on his agenda. Aaron makes an appointment to meet with Craig.

Note

1 First used in the comic strip "Pogo," by Walt Kelly, in the 1960s.

5

Weak Culture Miseries

Aaron thinks about his future with his struggling company as he wrestles with what kind of future awaits him. After all the frustrations, discussions, ups and downs, Aaron decides it is his turn to be completely committed. At this moment Aaron's decision means he needs to take responsibility for developing a strong safety department that can both serve and lead the needs of his company's safety, health and environmental issues. However, all of the above are not looking all that great at this moment in time. He considers the recent coaching he has received and how this can help him move forward with the daunting challenges that his commitment to stay, serve, and excel will require of himself and those who work with him.

Aaron's next gut check is not all that encouraging. The Doc talked to him about leaders, followers, and resisters. In a weak safety culture it seems: leaders go to more important jobs, followers follow them, resisters and poor performers are put in safety where they can be hidden. Indeed, this is Aaron's lot in the life as a safety manager in an organization which only gives lip service to safety. This too had been the Doc's experience with the Fortune 20 company trying to dig their way out of the poor safety culture hole.

Aaron pulls up his consultation notes and reflects on what the Doc refers to as a unit operation called excellent safety culture performance, but which safety societies often think of as one of a number of normal safety processes. As with any other unit operation/normal safety process, there is an ongoing dynamic of many moving parts:

- Each of the number of individual process steps must be done well.
- All these moving parts must fit together if an organization is to achieve anything that has a chance of putting an end to a culture of mediocrity.
- Any attitude of "good enough" or "close enough" is a death knell to achieving anything in which people of the organization want to commit to delivering world-class performance.

Delivering Safety Excellence: Engagement Culture at Every Level, First Edition. Michael M. Williamsen.
© 2021 John Wiley & Sons, Inc. Published 2021 by John Wiley & Sons, Inc.

What are the necessary steps to achieve safety culture excellence? Aaron noted extreme shortfalls in: management commitment, leadership involvement for hourly and salaried personnel, training, engagement, near miss, etc. ad infinitum. But which one(s) do you lead with? There is just too much to do with not sufficient resources, and everything needs to be done now.

There is the additional complication that over the years a weak culture seems to reward poor performing salaried employees by not requiring much in the way of anything which even remotely approaches excellence. As a result, the status quo momentum of "good enough" drags down one part of the organization after another. This type of vicious negative cycle is especially true in a company like the one where Aaron works. Aaron faces daily frustrations as ER (Employee Relations department) sets hurdles that take forever for Aaron to overcome as he tries to resolve the poor performance of the few bad apples who drag down the overall performance of the entire group. So where to start? Aaron decides to take on a challenge required for his success. With this decision he commits himself to successfully navigating all the ER and interpersonal challenges required to shed himself of his worst bad apple person. If the safety department is to move forward this is Aaron's personal "must solve" challenge.

> *Thus, Aaron returns to his decision of a commitment to getting rid of the worst apple. He must take the lead in doing whatever it takes to resolve his personnel issue. ER cannot be an excuse for Aaron in this matter. If Aaron is to run the race set before him, then he must personally be successful in getting over the hurdles as they come up.*

> *As this saga unfolds, in however long it takes, the second part of the weak safety culture strategy has Aaron beginning a process of mentoring the other few safety staff members in order to help them deliver better performance. Aaron plans out the needed steps for success. As a part of this improvement strategy he also decides how he will give them the needed personal appreciation/recognition for good performance as it occurs. Personal, frequent, sincere feedback for a job done well is a true asset in lifting the performance of an individual and an organization. Such reinforcing feedback is also a necessity for the team comprised of the various individuals who are keeping all the moving parts going in the correct directions. Aaron wishes he had been given this kind of reinforcing feedback, but no need to whine about past events. Therefore, he realizes that, "I must take responsibility for what I can do now to improve our lot. If it is to be, it is up to me; so suck it up buttercup. Here we go!"*

> *And then a significant surprise occurs when his best trainer, who quit because of the worst apple, comes back and tells Aaron that if he can get rid of the bad apple he would like to come back and help the department to improve the company's safety performance.*

6

Injury Plateau

Aaron digs into the many difficulties of building a successful case necessary to remove his poorly performing employee. As he does so there is welcome relief from the long, hard grind in his next task; building the skills of the safety department's other resources. These two very different tasks are both about building personal relationship skills which are necessary for a team to function well with positive reinforcement, adult correction and skill building. Aaron smiles as he realizes this part of the plan to improve safety is coming together. But then there is still the cloud of a just plain sick safety culture that rains down on all the employees every day. How to "eat" this next elephant in the room will be just as challenging, if not more so. After his many discussions with the Doc, Aaron recognizes that applying the classic safety tools will do little or nothing to improve safety performance at their RIF plateau of 10+. These standard safety approaches will not address the real issue of a very weak safety culture. He writes down the "usual suspects" and thinks through what the result would be if this was all he planned to do:

- *The OSHA compliance approach of necessity has a condition centric focus. Aaron's injury and incident information show that facility physical conditions (and not personnel actions) account for very few of the injuries. Furthermore, the compliance tool is always reactive to an event that has already occurred. What is needed is a proactive strategy which will prevent incidents from occurring. His compliance audits have strong scores, the basics are in place. But his organization's current safety culture management has no thought of improving the underlying culture, just keep doing the same things and hope to get better;" The beatings will stop when the safety performance improves"*
- *Aaron did not give high marks to the anemic data from the "program of the month" behavior based safety (BBS) which was in place before he took over. The project kicked off well, but then faded within a few months as all scores seemed to magically get to excellent, even though injuries continued to occur. Aaron picked up an*

Delivering Safety Excellence: Engagement Culture at Every Level, First Edition. Michael M. Williamsen.
© 2021 John Wiley & Sons, Inc. Published 2021 by John Wiley & Sons, Inc.

article the Doc sent him and reached the same revelation that was explained by the probability equation from Dr. Dan Petersen; the fewer observations that highlighted safety problems, the more observations are required to be statistically significant. Dr. Petersen's article provided statistical insights not normally considered by BBS users:

Limitations of Safety Observation Sampling

Formula for calculating effectiveness

$N = 4(1-P)/(Y^2)(P)$

where: N = total number of observations required, P = percentage of unsafe observations/actions, Y = desired accuracy.

If desired accuracy Y is 10% and unsafe actions P are 10%, then N = 3600 required observations.

With 1% unsafe actions P and 10% desired accuracy Y, then N = 39 600 observations to be required.

What are the manpower requirements required for 39 600 observations?

The BBS process plateaus before the injury rate declines to world-class low levels. The bottom line is that more and better tools are needed for improvement.[1]

As with quality improvement: you cannot inspect quality in, nor can you observe safety action levels enough to achieve excellence. You must fix the upstream processes that deliver the necessary downstream performance. The safety improvement culture must go beyond another observation and/or compliance program and into improving the safety culture. Such improvement must ultimately move beyond the basics which are of foundational importance, but only help to deliver world-class results. As Aaron searches for effective safety processes he finds that the literature does not have much in the way of other tools to use. The majority of the references emphasize reactive, fundamental government regulations and the soon plateaued observation program of the month. Aaron knows he has to go beyond this approach, but he is not all that certain of what can be used to engage the minds and actions of his people.

In order to answer this dilemma Aaron goes to his desk, picks up the business card, and makes the call. The Doc picks up and listens intently as Aaron explains his safety culture successes, failures, and struggles over the phone. When asked "What has really happened to cause all the injuries?" Aaron comments about the safety filter discussion they had during the interview process (Chapter 1). Aaron's organization, with a RIF of about 10, has plenty of holes in the various safety filters/barriers of management systems, working conditions, and human behaviors. The high RIF rate tells him that the holes in these safety barriers often line up resulting in injuries that frequently occur. Last year Aaron's company experienced 90 recordable injuries, of which 39 were lost time cases. This year when the holes lined up, a fatality occurred, in a classic example of H.W. Heinrich's controversial predictive model. What is needed is to plug these holes. Aaron and the Doc agree to carry on an in-depth conversation while on site. During this face-to-face conversation they will discuss how to significantly improve the safety culture of Aaron's company. In turn, this process will lead to the real on-site work necessary to improve the weak safety culture.

> *Aaron "sort of" feels good about this approach. Yes, moving beyond merely talking about safety improvement is the correct thing to do. However, he knows that he and his company are now about to head into the long, challenging work of actually accomplishing a significant safety culture improvement for the more than 1000 employees of the company. Moving from talk to walk is never easy, but is always a necessity if any lasting improvement is to occur. Status quo won't provide any improvement for his many friends at work who experience significant injuries each year. Once again, Aaron is committed to taking the next step in a marathon journey to safety culture excellence. There are no short cuts, there are no quick fixes. The real work is about to begin, and it won't be easy. It reminds him of the story about the person waking up from a dream in which he was a butterfly going from flower to flower. For a moment he didn't know if he had been dreaming about being a butterfly or if he was really a butterfly having a nightmare!*

Note

1 Safety Management, A Human Approach P. 349, 350, Dan Petersen, Aloray Inc. Publisher, Goshen, New York, 1988, ISBN 0-913690-12-0

7

A Brief Safety History

Before diving into a full court press that will improve his organization's safety performance, Aaron and the Doc decide to do an onsite walk around of a number of the departments and field operations of the company. He and the Doc talk about safety culture as they did before, only back then to a much lesser degree. The dialogue begins with a brief history of safety in 1911 and then continues with examples of a number of significant improvements since then. In his mental background, Aaron wonders if he really has time for this. He knows another injury has occurred at the company and fears the repercussions which always seem to come from an injured employee who voluntarily admits to an industrial injury, turns in the paperwork and prepares to run the gauntlet imposed on them by upper management. But the Doc is seldom available onsite, this is an opportunity that can't be dismissed. Aaron goes on the tour hoping that maybe this time he, and the injured employee, will get lucky and escape the gauntlet.

Back in 1911 our country had no formal safety emphasis. Injuries and fatalities were common as the United States was receiving 100 000 s of immigrants yearly. Few of these immigrants spoke English, all needed jobs, education levels were generally low and menial hard labor jobs were in abundance for all who would toil in their own process of starting a new life. These immigrants hungered for a new life which would allow them to bring over their family members from wherever they came and fulfill their own personal dream of prosperity. And then there was a tragic trigger event in one of the many sweatshops that abounded, all of which had little or no safety considerations whatsoever. In 1911 a fire at the Triangle Shirt-waist Factory in New York killed 146 people. There were huge public outcries, which ultimately lead to the beginning of safety laws, their associated penalties, and many safety technology improvements. The tragedy was also the launch of a safety technical group which eventually became The American Society of Safety Professionals (ASSP).

Delivering Safety Excellence: Engagement Culture at Every Level, First Edition. Michael M. Williamsen.
© 2021 John Wiley & Sons, Inc. Published 2021 by John Wiley & Sons, Inc.

Indeed, the Doc's family background in the United States also began in this New York environment in this era. The Doc's grandfather, Hermanus, arrived at Ellis Island, New York in 1905 with one steamer ship trunk that contained all his earthly belongings. He had an all-consuming desire to begin a new life and make enough money to bring over the woman he was pledged to marry. He spoke no English and had no formal education. However, he did have: a strong work ethic, his burning desire, a solid future dream (vision), and a small family of relatives who sponsored him in Nebraska. The Doc mentioned that he has often told his children that their great grandfather came over to America to fulfill his dreams, but in actuality he engendered multiple generations of descendants who are still fulfilling their own American dreams.

In this early era of safety there was no background safety material or safety technology. If a worker got injured and could not do the daily hard labor work, they would have no money with which to eat, or fulfill their family dreams. People were responsible for their own safety and working conditions were atrocious by our current standards. In many cases, each step along the way over the last 100+ years in the safety technology and culture continuum was the result of trial and error, not from a scientific solution process. However, progress was made by reacting to conditions and developing sustainable fixes for the many hazards, e.g. guarding, ladders, stairs, etc.

A few companies in dangerous industries, e.g. DuPont and Dow Chemical, recognized the necessity of safety excellence if they were to stay in business. These companies set in place policies, procedures, and reactive standards around dangers which existed in their work places (fire, explosion, power isolation, etc.). Theirs was a step-by-step process which intensely and thoroughly addressed one issue at a time and then moved on to the next potential safety disaster.

As time progressed, some scientifically minded analysts begin to look at accident/injury causation: originally Herbert Wagner (H.W.) Heinrich with his Workers Compensation (WC) analysis, then Frank Bird and a host of others as time progressed. H. W. Heinrich changed the world of safety fundamentals forever with his pioneering work in the 1930s. Keep in mind there is still significant controversy among safety professionals about how statistically relevant Heinrich's research was. The statistical proof of his incident number estimates is not the point of this history discussion. Rather, it is one of his concepts that continues to cause many safety professionals to think of his accident triangle (pyramid, Figure 7.1), a concept with which we are all familiar.

The Heinrich theory states that so many near misses/close calls lead to a lesser number of first aid injuries and thence onward, through his logic, to recordable injuries and ending in the inevitability of a fatality. This inevitability of disaster has always bothered the Doc. The Heinrich model seems to imply that if people cross railroad tracks too many times they will die, or drive to work too many times,

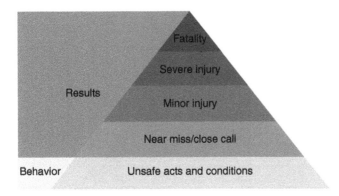

Figure 7.1 Heinrich accident pyramid.

or something else like that. Those of us who are not fatalists often have pondered about what could enable the industrial work place to overcome this fatalistic teaching.

In answer to this conundrum, Dr. Petersen's and the Doc's work with companies and individuals, which have done both well and poorly in safety, has always led to individual behaviors as a foundational key to low injury rates. That is, as long as workplace conditions, training and safety standards are also addressed. However, in many cases, the "behavior foundation" appears to be made of sand. Often it has not given a firm foundation on which to build a zero incident/at risk activities safety culture. So how does one attack this foundation of the Heinrich triangle? Years of thought and effort in this area have led to a whole different kind of sub-foundation fundamentals as shown below.

The solution occurs when this new safety triangle (pyramid) is built on the "solid rock foundation (Figure 7.2)" of excellent compliance-related fundamentals and error proofed fundamental safety processes, e.g. incident investigation, excellent communication, sound supervisor leadership and performance, etc. In turn, when practicing a culture of excellent behaviors and actions, an organization's leadership can significantly limit (reduce) the base of unsafe/improper acts and conditions which seem to lead to 80+ % of the injuries in Heinrich's pyramid. In Dr. Petersen's and the Doc's experiences, each time they were involved with groups that rebuilt the foundation, the results were similarly excellent.

Over the years Dr. Petersen discovered that the following sub-foundation fundamentals also contributed to significantly reduced injuries:

- Visible upper management leadership in safety.
- Noticeable involvement of middle management.
- Focused supervisory performance.
- Active participation of hourly employees.
- Training that both teaches and reinforces this type of foundational excellence.

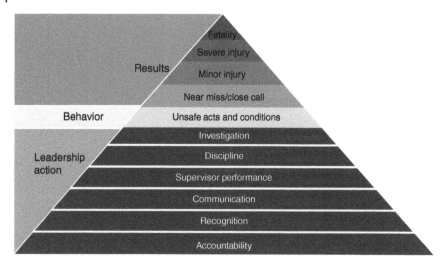

Figure 7.2 Enhanced accident pyramid.

The above principles tie in very well with the quality function's 6-sigma initiative for significant improvement, DMAIC (Define, Measure, Act, Improve, Control). Only in the safety case it is:

- Define the correct behaviors which help eliminate unsafe acts and injuries.
- Train all personnel in the behaviors which will improve both individual's and the overall safety culture.
- Provide the necessary resources to improve safety performance.
- Measure that people are indeed doing these correct behaviors in a systematic quality way.
- Give recognition to individuals and their work groups as they accomplish these correct behaviors.

The key then is not to focus solely on compliance, or reward "acceptable injury levels/goals," or on any other single tactic approach. The key is to concentrate on the foundational fundamentals which help to eliminate the unsafe activities/behaviors. This approach, in turn, will move an organization out of the injury probability of the Heinrich inevitability triangle. A world-class safety organization must error proof the fundamental safety processes while additionally living viable safety accountabilities at each level of the organization. In so doing, their people then engage in a daily and regular practice of the upstream activities which deliver excellent downstream performance.

The question about what exactly are these *fundamental safety processes* was the object of a 10-year study by Dr. Dan Petersen and Dr. Chuck Bailey (see Appendix B). They worked together on an in-depth study of railroad industry injuries and cultures. As a result, they developed a safety culture analysis diagnostic which points out safety culture weaknesses at three critical levels of an organization (frontline labor, frontline supervision, upper management). The in-depth analysis of these two safety culture pioneer researchers went beyond the typical reacting to conditions by focusing on upstream processes which, when done well, significantly improved downstream safety performance. That is, as long as the processes were linked to appropriate safety accountabilities at each level of the organization. As they wrapped up their study and the diagnostic, they concluded that there were 20 of these fundamental upstream safety processes (which they called categories). All of them were necessary for an organization to significantly improve downstream safety performance.

Their safety perception survey had its beginnings in the late 1970s when Dr. Dan Petersen and Dr. Charles (Chuck) Bailey teamed to study what really made a difference in safety performance. The research took place in two parts. The first focus was on what "conventional safety wisdom" said made a difference, i.e. procedures, engineered solutions, OSHA regulations, and the like. This five-year effort showed that "The most commonly used criteria in standard safety program audits are poor measures of program effectiveness." "This lack of significant impact exists because the added attention to safety and health generally has been focused primarily on the requirements of regulations instead of on the reduction of risks, in large part because the regulatory system focuses more on means than ends in the control of potential hazards" (see Appendix B).

The first study also took input from safety professionals on processes and questions which they thought would make a difference in safety performance that were observable. As a part of this discovery effort, approximately 1500 potential survey questions were distilled down to 73 questions (see Appendix B) which the study showed to have "a statistically significant response." The validation approach compared question responses from a poor safety record (high injury rate) organization to those of a good safety record organization (lesser injury rate). All questions that had no significant statistical difference in response between these two very different safety record organizations were eliminated from the survey. These validated questions were then mapped to 20 safety processes (Figure 7.3) that, when existent in a safety culture, should consistently result in observably better worker safe practices. These observably relevant questions and their 20 "management systems" were then retested in a second phase of the 10-year study.

After reviewing the data from the first study the second multi-year "verification study" focused on positive reinforcement of correct safety activities and "the quality of the 20 management systems which have an effect on human behavior relating to safety." They found that; "The most successful safety programs are those which recognize and deal effectively with employee and supervisor behavior and attitudes which affect safety."

❑ Attitude toward safety   ❑ Management credibility
❑ Awareness programs   ❑ New employees
❑ Communication   ❑ Operating procedures
❑ Discipline   ❑ Quality of supervision
❑ Employee training   ❑ Recognition for performance
❑ Goals of safety performance   ❑ Safety climate
❑ Hazard correction   ❑ Safety contacts
❑ Incident analysis   ❑ Substance abuse
❑ Inspections   ❑ Supervisor training
❑ Involvement of employees   ❑ Support for safety

Figure 7.3 Safety management categories.

Additionally, the research team began to engage in problem-solving efforts around the weaker scoring of the 20 processes which the survey showed to exist. In this phase of the study question responses from another set of good and poor safety record organizations were once again compared to confirm the validity of high question scores to good safety performance (low injury rate). The end result was a strong statistically valid correlation between:

- These 20 high scoring processes.
- An organization's improvement efforts to low scoring processes.
- The training of supervisors in these improved safety management systems.

The end result was the consequent lower percent of observable unsafe worker behaviors and subsequent lower injury rates.

And thus the conclusion that, when present at higher percentages, these 73 safety culture questions and the associated processes they map to, were statistically valid indicators of the "human factor" and associated worker activities, and therefore a culture of better safety performance (lower injury rates). Or as their study conclusions stated "Application of 20 category survey techniques developed by this study provides a reliable measure of safety program effectiveness."

Building a safety culture is a continuous process that needs deep roots to stand on its own, yield fruit and become part of the environment

Figure 7.4 Building a safety culture.

Their pioneering research efforts set the stage and foundation for an in-depth dive into going beyond reacting to conditions and into what it takes to develop safety culture excellence. When these seeds of their 10-year study are planted, cultivated, watered, and fertilized, i.e. tending the fields, it becomes possible to harvest a crop (culture) of safety excellence (Figure 7.4).

However, using this same farming/agriculture analogy (Source: Steven Simon, PhD), it also apparent that an excellent safety culture must continue to tend the fields they have so carefully prepared. If an organization neglects a culture of excellence in the accountabilities that help to deliver high performance, things go back to the unmanaged state of weeds, poor crops, and death of the crop/fruit-producing plants. In a safety culture this means incidents, injuries, and fatalities return when a weak safety culture returns and becomes prevalent.

> *Aaron's one day walk and talk with the Doc had him introducing the Doc to much of the organization and its people. The time together also introduced Aaron to the more than 100 year progression of safety culture. This progression has now led to a focus on the development of fundamental safety upstream processes which all employees live as a part of their practical safety accountabilities.*

> *Ok, but how does our weak safety culture make this significant shift from WOW (Worst Of the Worst) to BOB (Best Of the Best)? This transition is especially difficult when Aaron's management resists any change, or improvements, from the compliance and regulations models.*

> *As their day long tour and talk comes to an end Aarons' cell phone rings. His boss has another injury problem for Aaron to investigate, and the gauntlet looks to be about to take place. Aaron has begun to understand that the real issue is a lack of the organization's efforts to improve the fundamental safety processes and implement necessary safety accountabilities for anybody, other than Aaron, who seems trapped in the role of a safety cop. Aaron bids the Doc goodnight and heads to his boss' office.*

8

Beyond Accident Reaction

Aaron goes to his boss' office to review the latest injury. He anticipates a difficult meeting, what other kind has there ever been with this guy? However, after the daylong safety culture discussion with the Doc, Aaron decides on a different tactic from his usual tail between the legs, dog whopped approach. Maybe, just maybe, he can use reason and intelligence to turn the tables on what he has experienced over the last few, of what seems like many, months. So, Aaron kicks off the meeting by complaining about the lack of management support that has led to:

- *The latest injury*
- *No improvement in quality or quantity of the safety department staff*
- *No engagement of the frontline employees in helping to solve the problems*
- *No support to develop and live proactive upstream focused safety accountabilities across the organization*

In return, his boss looks at the injury report and concludes the employee was careless and needs to be punished. He instructs Aaron to turn this over to ER. After all, the recent internal OSHA audit showed the company was in decent shape with respect to the regulations. The real issues his boss believes are poor employee work habits and bad attitudes, and these won't be improved by Aaron's wish list of non OSHA objectives. Additionally, there have always been safety and performance problems when storm conditions occur. Since the forecast is for a severe storm across the weekend Aaron needs to cancel weekend time off and prepare to call in contract labor just in case. "So, Aaron, you must start doing the real work of our company."

The result of Aaron's attempt to reverse the tables sure feels like another dog whipping. Oh well, another day in paradise for the safety manger. However, rather than giving up to a poor performance safety culture, this latest whipping reinforces Aaron's and the Doc's belief that Aaron's company's lack of safety

Delivering Safety Excellence: Engagement Culture at Every Level, First Edition. Michael M. Williamsen.
© 2021 John Wiley & Sons, Inc. Published 2021 by John Wiley & Sons, Inc.

progress is more an indicator of an overall government, academic and indus-try leadership's resistance to change old paradigms. The result is that while background, observation and culture work is going on, annually the United States is still experiencing 1000s of fatalities and 100 000s of injuries in the industrial workplaces across the nation. The frustration with this lack of sig-nificant progress in reducing the carnage led many to count on government to help solve the problems with a new set of laws, and OSHA was born. But, this regulatory approach, though fundamentally important to any safety culture, is not delivering anything near to that of the zero incident performance which is needed.

As the next day begins, Aaron and the Doc review some more history. Before Aaron's current time (in the barrel?) all that seemed to exist were materials from a few companies which had better injury records. Examples of this are the policies and procedures of reacting to conditions by companies like DuPont and Dow Chemical. Their brief paperwork policies and procedures were handed over to government lawyers and a few thousand pages of CFR 1910 (Library of Congress Federal Register part 1910) regulations were born along with a cadre of govern-ment inspectors who review workplaces, write reports, and issue monetary fines for noncompliance. The inspectors typically have limited industrial experience and are further complicated by many branches of the OSHA law which require in-depth expertise in complicated technical arenas like electricity, guarding, PEL (Personal Exposure Limits), fire prevention, etc. Additionally, there always seems to be significant resistance to inspections and resultant fines by a wide range of industries. This approach has delivered a less than satisfactory result of a government industry battleground being staked out. There is some good news; some progress is occurring, but lasting and significant injury reductions are slow to occur.

The regulatory battles seem endless, and yet some people began working on other models in hopes of finding the "silver bullet" which would solve the never-ending "incident reaction cycle (Figure 8.1)."

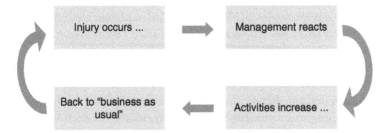

Figure 8.1 Accident reaction cycle.

Among the silver bullets which were hoped to significantly solve the injury reality are approaches the safety technology history has added like:

- BBS (behavior based safety) being developed and marketed.
- Psychology being studied and applied.
- Trinkets and trash (monetary and/or other tactical rewards) being marketed for low injury rates.

All of these changes seem to show some progress, but these tools appear to deliver no lasting, significant, sustainable improvement shift. The colleges which teach safety continue to significantly focus on regulatory compliance. The graduates who enter industry implement what they were taught and what is reinforced by OSHA. As a result, it seems the safety incident progress of industrial organizations stalls (or plateaus) no matter how hard these fundamentals are pushed. This is symptomatic of the dominant culture efforts to press forward with a fundamentally reactive approach to safety. All the while, WC (Workers Compensation) costs escalate as a shift in the workforce makeup only very slowly begins to affect injury statistics. Meanwhile the increase of raw insurance and medical treatment costs continue to escalate.

The Doc waxes eloquently about how worldwide there has been a significant advance of multiple scientific and industrial technologies, yet the educational system is experiencing a degradation of the skills necessary to design, manufacture, and maintain the technologically advanced workplace. In turn, this shortfall helps to deliver a huge gap in performance. The required labor force is no longer served by the old model of low education, low skill, menial labor. Company upon company, industry upon industry, now scrambles to get the more highly skilled labor it needs to effectively compete in an ever more global society. However, the segment of the work population which gets injured remains those people on the frontline, but now they are the hardest to hire, train, and retain. Day labor in the global competitive environment of heavy industry is no longer a significant performance driver. As cost and performance pressures ever mount, more and more companies figure out a huge cost reduction and potential competitive advantage are delivered by eliminating injuries at the frontline. These high tech workers deliver what it takes to pay the bills and salaries of all those on the organization chart who are above them. Many leaders in industry have begun actively pushing for a zero incident (zero at risk activity) culture which preserves their critical living asset base.

The Doc talks about how the need for quality excellence began to occur similarly, but somewhat ahead of this new safety culture paradigm time. The zero defect quality necessity and resultant culture of excellence in quality control also had its thought leaders: Joseph M. Juran, Phillip B. Crosby, and Dr. W. Edwards Deming. They noticed a lack of usefulness for the lagging statistics similar to what was to occur later on in safety. The incident reaction cycle (Figure 8.1) is just a modification of the quality defect reaction cycle these pioneers in quality had to contend with. Percentages of scrap and rework were only indicators of process failures and a cultural malaise which accepted a reasonable failure rate, i.e. the industry standard for scrap and rework. Instead of counting what they did not want to occur, these quality leadership pioneers began to error proof the upstream processes which delivered the downstream results of reduced failure.

They had to go beyond fixing manufacturing equipment capabilities. They had to:

- Improve detection capabilities.
- Improve frontline worker skills and capabilities.
- Develop and live personal accountabilities at every level of the organization, which helped to eliminate quality defects *before* these undesirable events occurred.

In order to achieve their goals they had to develop a quality culture which engaged in the relentless pursuit of zero errors every day, in every way. To do this, they began to engage the people who knew the problems most intimately, the frontline workers. In time, the quality culture became a kaizen (continuous improvement) mindset. This active kaizen culture on a regular basis engaged the brain power of small teams that were comprised of both the hourly and salaried employees. Or as Dr. Deming is rumored to have said: "You pay for the body and the mind comes for free." Using this approach, the globally dominant organizations improved quality, performance, longevity, and sustainability well beyond what was thought possible in the 1960s. Their cultures (what the workforce and organization did without even thinking about it) relentlessly pursued and delivered a zero incident quality culture. The standard they aimed for became 6 sigma performance (three measurable defects in one million final products).

This brings to mind a story the Doc's Papa told him about his experiences of buying an automobile in the 1950–1960s. As Papa talked with the various dealerships, which at that time sold a much larger number of America-made brands and models than we have today, he got the "best available deal" that was referred to as the "Gold Bond Guarantee." And that best deal was a money-back guarantee on his purchase of a new car for the first 90 days or 3000 miles, whichever came first. Wow, have quality, competition, and technology ever changed the concepts aligned with personal vehicle excellence!

Consider then the concept of beginning with the end in mind: What is the difference in approaches to achieving global excellence in:

- Productivity
- Quality
- Customer Service
- Safety

There is No Difference!

Building A Zero Incident Safety Culture

A Long Term Process, like building a winning team . . . . .

- Perfect the fundamentals and the "plays"
- Develop the individual and team skills
- Engage all team members
- Focus on excellence every day in all that you and your team do

Or *like a growing tree, it takes three to seven years to. . .*

- Develop deep roots
- Stand on its own
- Yield fruit
- Become a permanent part of the environment

All of the above are about developing a long-term organization that maintains and lives a sustainable safety excellence commitment. It comes down to delivering impeccable quality (and safety) performance or regressing to the non acceptable alternative of going out of business as competitors keep pushing the quality (and safety) envelopes.

Drs. Dan Petersen and Charles Bailey, and a very few others, began considering and developing a culture of accountability. They desired a culture where people across the organization from the C level (highest leadership/management level of the organization) to the frontline engaged in meaningful proactive accountabilities, which in turn would reduce the possibility of injuries. Developing and delivering what this new approach promised was a significant challenge in the existing culture: regulatory compliance only, hiding of injuries, the 800 pound operations gorilla of cost, quality and customer service, and a subtle shifting of worker technology necessities. All these items coupled with no known, proven means of achieving a zero at risk activity/incident culture had the academics writing research papers which filled volumes, but delivered no real, substantial solutions (to what seems to be basic human nature realities). As in the quality revolution, the academics needed to partner with industrial organizations which could help them both achieve and document 6 sigma safety performance (three safety risks per million hours of work). This innovative safety culture vision is more than four orders of magnitude better than the current industry average RIF[1] of 3+ medical injuries per 100 employee effort years (3% of the total workforce being injured seriously enough to require medical attention).

Aaron and the Doc discuss how the current culture at Aaron's organization certainly lives the incident/accident reaction cycle. When an injury occurs, management demands actions that will prevent a similar injury in the future.

And for a while the injury rate goes down. However, since the real empha-sis of upper management is in on the big three of cost, quality and customer service, soon the safety lessons are forgotten and then inevitably the next error, injury, occurs. In turn, this approach initiates the same accident reaction cycle of insanity. The only way out of this dynamic injury trap is found in step three of the incident reaction cycle (Figure 8.1); living activities (safety accountabil-ities) at all levels of organizations, which help to prevent injuries.

Aaron nods his head and asks the salient question: "You have told me all about the what. When are you going to get to the how? How do we get out of the academic philosophizing and into the concrete necessities of safety per-formance/culture excellence?"

Note

1 *RIF Recordable Injury Frequency ([number of injuries requiring medical/doctor treatment] × 200 000) divided by total hours worked.*

Part II

9

Safety Culture Beginnings

Aaron leans back, closes his eyes and thinks about how he is going to handle the "same stuff, different day" culture which is rampant in the company. He makes a "sucks list" which includes: all the various people groups at each other's throats, being stuck with only the same old reactive tools and approaches with no one seemingly interested in pitching in to help improve their sick culture, and no assistance or coaching on how to handle all this "stuff", etc., etc., ... With a gloom and doom face Aaron trudges out to meet with the Doc. Maybe today some light will shine through. Aaron could sure use a ray of light or two.

The Doc senses Aaron's depression and shares personal story details of when he was director of operations for a facility that produced military ammunition and also demilitarized (demil) dangerous obsolete ordnance. The end of each calendar quarter and year end were always at risk with respect to safety as there was tremendous pressure to produce and demil every last round so the period's financials looked as good as possible. During these periods, overtime hours were huge. Then, once production was completed, there was a significant layoff, especially just before Christmas. Once contracts were re-established in the next year people were rehired for the chaos of production in the next quarters. Safety was not even a consideration for any of the operations when the Doc took over responsibility.

Their moribund safety approach was delivering a 142 severity rate[1] for this 500-person operation and a history of severe authoritarian management. It was obvious to all frontline employees that people were merely viewed as numbers and safety was talked about, but never really practiced. Indeed, the organization's nine on-site professional safety staff resources concentrated on a punitive approach as a means of trying to improve safety. Aaron relates to this approach very well, except for the last bit about **NINE safety resources!** and management, not the safety department, issuing punishment. Aaron cuts to the chase and asks the Doc;

Delivering Safety Excellence: Engagement Culture at Every Level, First Edition. Michael M. Williamsen.
© 2021 John Wiley & Sons, Inc. Published 2021 by John Wiley & Sons, Inc.

"So how did you get out of this same miserable hole I find myself in? And how about skipping the philosophizing and giving me the facts!"

The Doc smiled and then explained that some amount of background detail was needed. He related about how one of the more dangerous operations in ammunition manufacturing was inserting explosive primers into the cartridges. The highly explosive primers were always to be handled without metal tools, because of the fear of static electricity setting off the highly explosive (1.1) energetics. As the Doc drove into the facility early one Friday morning there were ambulances outside the ammunition manufacturing cell bay. The guard told him there had been an explosion with two men injured. The Doc quickly suited up and went to the primer loading facility and talked to the crew and their supervisor. Yes, two mechanics had been injured in a flash explosion and were being taken to the hospital. After arriving at the hospital the Doc talked to the emergency doctor and was assured that neither man had significant injuries, but each did have facial flash burns which, in the medical doctor's opinion, were not serious. He felt they were very fortunate because they had been wearing the required PPE (Personal Protective Equipment). He thought one man should stay overnight for observation and the other could be released immediately. The Doc then visited and talked with both men who were visibly in a state of shock, but otherwise seemed fully functional. After this, he then talked to the doctor and asked him to hold them both across the weekend unless they wanted to go home and were okayed to do so by the hospital. There was a comment made by the physician that this process would make them both be considered as lost time injuries by OSHA rules and that he knew the company the Doc worked for did not like that. The Doc stepped up and assured the physician who was in charge of the ammo facility and that he (the Doc) wanted the best for the men. The Doc was not concerned about whether or not their injuries became avoidable lost time injuries. The Doc then went out and visited with the gathered family members and assured them of what the physician had said and suggested they could go make a short visit with the two injured men they loved and cared about.

Both men easily returned to full duty work the next week while a small team of facility staff members did a detailed incident investigation. From the site evidence, it appeared one of the injured had tried to clear a primer jam with a metal pry tool, instead of the wooden tongue depressors which were supposed to be used for clearing such a jam. This process then likely led to a static electricity charge which set off the very volatile primer material. The safety staff and other long time staff recommended DA (Disciplinary Action) which the Doc immediately negated; there had already been more suffering than anyone wanted. A lesson had been learned, and a DA would only make things worse in his opinion. Instead, the Doc commissioned a small kaizen/continuous improvement team to take a deep dive into why these primer jams occurred and then fix the root causes. Within a month this personally motivated team of hourly, supervision and engineering volunteers then solved the primer jamming problems. These operational defects apparently had been an issue for a very long time, but the problem had been tolerated, instead of the root causes being solved.

About four months later, an OSHA field representative showed up to evaluate the incident. The Doc was surprised, but sat down with him to discuss the representative's mission and see what the plant staff could do to facilitate this officer's investigation. During the talk time, the OSHA officer mentioned he had responsibility for 39 counties in the same state and as such, only had time to work on fatalities. "So why are you here for what is very far removed from a fatality?" the Doc asked. The return comment was that in his eight years with OSHA the officer had never been a part of an investigation with which he had any previous experience. However, as an ex-army ordnance disposal unit member, he was looking forward to at last being involved with an issue he knew something about. The Doc made sure plant personnel fully cooperated with the OSHA officer's desires in the investigation. Some months later the OSHA report was received stating the officer had come to the same conclusions as the plant team, and no fines had been levied. The Doc congratulated his team for doing a very good job of analyzing the event and then delivering solid root cause solutions which eliminated the causes of such future primer jams. Additionally, there was the lesson that the year and a half of teamwork efforts on all the condition and training issues had led to no findings by the OSHA officer. Condition issues are important; these foundational safety necessities must be resolved; your people can do so if you support and reinforce them in the process.

As a part of this injury event, and others, it became ever clearer to the Doc that the available safety technologies were almost all reactive with the following sequence; injury rates, incident investigation, observation programs, near miss/close call investigations, compliance fines, etc. Added to this cycle was the fact that the available solution tools were very weak, especially the use of tired and insignificant safety teaching videos. Things did not look any better to the Doc when he was also faced with the poor leadership capabilities of safety staff members. It was no surprise that the end result was a weak, stagnant safety culture. This culture plodded along, never seeming to deliver anything but mediocrity. Indeed this approach seemed to be the norm for many American safety cultures. This continued to occur even though over the years OSHA had added some new laws, after protracted legal battles with industry. OSHA also tried the VPP[2] approach, which awarded honors to those organizations which followed the rules and were able to meet their industry classification standard (average) for injury rates. The VPP approach seemed to focus primarily on compliance with OSHA regulations. Additionally, VPP offered STAR (exceptional performance) status for an organization's achieving an average injury rate in their industry.

Rewarding average (mediocre) performance is not a high enough bar for a group to achieve a zero incident/zero at risk activity safety culture.

The ammo plant practiced this regulations-based, fundamental approach on a daily basis with little or no improvement in safety performance. The deeper problems were really more the result of a poor safety culture which was evident by weak leadership at all levels. Necessary regulatory compliance was adequate, but this one-step approach is not good enough for obtaining excellence. It enabled and reinforced a "get 'er done now" production culture which did not engage people to solve the root causes of incidents. This weak safety culture did not reinforce leadership's and employee's need to be fundamentally in charge of protecting themselves when it came to safety both on and off the job.

The Doc mentioned that a cursory review of national safety data reveals that, statistically speaking, there was no significant improvement in injury rates within the status quo safety culture. With the exception of a few companies, good and bad, industry had plateaued when it came to injury performance. Schools continued to teach the reactive compliance technology in their safety curricula. Industry continued to battle the restrictive regulatory rules and, in so doing, injured more or less the same number of their human assets throughout time. Unions desired better safety performance, but were stuck in the same plateau paradigms which restricted the rest of the players. Academics continued to run studies and write papers to no seeming avail. Year after year the fundamental safety technologies delivered pretty much the same results. It seemed the process was stuck in Einstein's definition of insanity: doing the same thing and expecting different results.

In the academic safety community, as a result of his research results, Dr. Dan Petersen began to push the concept that we must go beyond reacting, counting what we do not want to occur, while additionally counting on (hoping) government entities to solve our problems. Petersen's research changed to a focus on improving the stagnant safety culture by implementing proactive safety accountabilities:

- Industry had been able to improve their cost, quality, and customer performance by engaging cross functional personnel in continuous improvement team solutions.
- After which they required proactive, focused daily accountabilities in manufacturing at all levels of the organization.
- Then why could not this this same approach work for safety as well?

Dr. Petersen believed that viable personal safety accountabilities throughout all levels of an organization would improve the downstream safety indicators. His opinion was that the natural extension of this approach would lead to a proactive culture of safety excellence. However, this was new technology in safety which

❑ Top management is visibly committed.
❑ Middle management is actively involved.
❑ Frontline supervision is performance-focused.
❑ Employees are actively participating.
❑ System is flexible to accommodate the culture.
❑ Safety system is positively perceived by the workforce.

- Dan Petersen Ed.D.

Figure 9.1   Six criteria for safety excellence.

did not seem to exist in industry, and thus he could only talk about a best guess as to what were the correct activities. His thesis was that organizations must build an intentional culture of proactive accountabilities at all levels which help prevent the possibility of errors (incidents and injuries). Petersen believed these safety accountabilities must be focused on upstream activities which deliver downstream results, just like those found and delivered in operations performance excellence models. He passionately felt that only in this way could an organization get well beyond the plateau being delivered by the reactive model.

Petersen's pursuit of his ideas led to his being thought of as one of the great safety pioneers of the last six decades. His focus was consistently on developing a viable safety culture which lived safety accountabilities at all levels of the organization. As a result of a significant amount of industrial frontline research, Dr. Dan Petersen struck upon "Six Criteria of Safety Excellence" (Figure 9.1). Years later, organizations fully utilizing his Six Criteria of Safety Excellence continue to be among the leaders in safety performance. These criteria are:

- **Upper management is visibly committed** to safety excellence. Visible means physically present within the workforce at the frontline when it comes to safety. Bulletin board postings, emails, and the like do not count; visible and credible physical presence with respect to safety is required. In most organizations it is difficult to pry executives away from their cost, quality, and customer service responsibilities and have them be visible in the workplace with respect to safety. In order for executives' credible field presence to be accomplishable, there is a need for realistic, value-added roles, responsibilities, and associated activities.
- **Middle management is actively involved** with safety. Middle management can be counted on to be present and proactive when it comes to the operations overwhelming attention to only (or as the frontline personnel call it "the gorilla of") cost, quality, and customer service. In fact, their pay and continued employment require that these factors be addressed when it comes to operations performance. So why not use the same approach in safety? One of industries' current realities is that there are far fewer middle managers than in years past.

This fact makes their active presence on a regular basis within the workface even more of a challenge to accomplish. Once again, practical roles, responsibilities, and activities provide guidance for these important people to make themselves known, and respected, in safety where it counts most, on the frontline.

- **Frontline supervision is forced to focus on performance indicators that deliver excellence** with respect to safety. As ever, the operations standards of cost, quality, and customer service demand this kind of detailed performance focus. So why not use the same approach in safety? In fact Dr. Petersen believed frontline supervision leadership was the most important key performance indicator in safety performance. Frontline leadership delivers performance, good or bad. Dr. Dan would often say "What gets measured is what gets done. And what gets rewarded is what gets done first." However, the issue here for the shortfall in safety was a lack of proactive accountabilities for hourly, supervision and upper management. Mostly, all that would be said with respect to safety was, "It is dangerous out there. You need to be careful and follow the rules." Clearly, this weak approach to delivering safety culture excellence at the frontline was woefully inadequate.

  Focused Supervisor Performance is another key attribute for excellent safety performance. Supervisors have very detailed and specific accountabilities for the cost, quality and customer service cultures found in the typical operations cultures. In safety, this level of detail and daily accountability is often lacking. Once supervisors get on the right detailed safety accountability track, their safety performance and that of their crew improve remarkably.

- **Active hourly involvement** in safety. An inadequate performance standard does not deliver any kind of excellent safety performance. There is no active hourly involvement in "It is dangerous out there. You need to be careful and follow the rules." What is totally missing in a reactive safety culture is the revolutionary performance delivered in excellent quality cultures by an active kaizen engagement of hourly employees to fix the problems they previously just tolerated. Active hourly participation, without a doubt, is of major importance. The hourly employees are the ones who deliver performance with respect to cost, quality, and customer service; why not safety too? They are used to accountabilities for everything except safety. Using this accountability criterion makes a lot of sense, and it works very well.

- **Flexibility**. Excellent performance for anything does not exist in a one-size-fits-all culture. A reactive safety culture still uses only a very limited, one-approach-fits-all, paradigm. One size does not fit all safety organizations and departments, even though a reactive safety culture often tries to force a very limited success probability, i.e. regulations and observations approach on all that exists. This approach just does not work. Appropriate departmental safety flexibilities that include items such as personal development, coaching, engagement, and recognition are a necessity.

- **The workforce has a positive perception** with respect to safety. Employees know truth, trust, credibility; and they can immediately spot anything phony. When polled in a nonthreatening manner, the frontline workforce can be counted on to deliver a sobering evaluation of what management's performance is truly perceived to be (and as the saying goes, perception equals reality). When positive perception is lacking at the frontline, one or more of the other five criteria of safety excellence are sick. This is the feedback loop in safety excellence. We measure our employees' perceptions and issues throughout the organization, and then engage cross functional teams in the work of developing and testing innovative safety-related solutions to the problem areas. If there is no effective feedback mechanism an organization quickly stagnates, and then deteriorates.

An organization needs to identify what needs to happen to improve a safety culture. It must do this before it is able to make the leap from "talk about" to actually changing culture. This kind of active, visible involvement is an important secret to attaining a sustainable culture of true safety excellence. The Six Criteria of Safety Excellence is an effective test for safety initiatives. Aaron realizes the application of the Six Criteria of Safety Excellence is a true shortfall within his company. None of these visible elements are a part of his organization's safety processes.

The Doc presses in about how Dr. Petersen's Six Criteria of Safety Excellence could just as well be viewed as the six criteria of operations excellence. Excellent organizations consistently deliver: visible upper management commitment, active middle manager participation, focused supervisor performance, active hourly participation, flexibility, and a workforce that is positive about their workplace and resulting production culture. Why not develop a culture that does the same in safety?

> At this point the Doc discusses with Aaron that, back when Dr. Petersen worked with the Doc, there was something missing with Dan's two pronged focus on; "what your culture truly is" and "what effective safety accountabilities are regularly and sustainably practiced." The missing element, which is present in excellent operations cultures, is the active engagement in a robust solution approach across all levels in the organization. The kaizen/continuous improvement engagement lessons of the quality culture revolution lead by Juran, Crosby and Deming seemed to be the missing element in delivering safety excellence. How could safety culture use these three quality pioneers' paradigms about:
>
> - The truth that upstream processes deliver downstream results

- *The need to stop counting and fixing scrap*
- *The necessity of fixing the upstream processes*
- *The revolutionary concept that you pay for the body and the mind comes for free, i.e., if we don't engage the intellectual/problem solving capabilities of the people who are doing the work, we are only getting a fraction of their abilities to deliver our many needs*
- *The strong and sustainable culture of the relentless pursuit of zero mistakes*
- *The kaizen culture of small changes forever*

The answer, based on the Doc's field turnaround experiences, is that a strong sustainable safety culture requires the engagement of people, at all levels of the organization, in 6 sigma kaizen approaches to safety improvement. There is a need for a safety performance culture which progresses beyond Dr. Dan's two step academic process of "I'll tell you the problem areas and you figure out what to do about them." A light goes on for Aaron about how to improve his safety culture. Once again, the drive home has Aaron in deep thought.

Notes

1 **Severity rate** is a safety metric used to measure how critical or serious the injuries and illnesses sustained in a period of time were by using the number of lost days (on average) per accident as a proxy for **severity**.
2 Voluntary Protection Programs is an Occupational Safety and Health Administration initiative that encourages private industry and federal agencies to prevent workplace injuries and illnesses through hazard prevention and control, worksite analysis, training; and cooperation between management and workers.

10

More Safety Culture

> *Aaron arrives on his usual 6 am schedule at his office the next morning. With the Doc here and the usual daily whirlwind of tasks and emergencies Aaron is slipping into sleep deprivation. Along with this fatigue comes the tendency to feel overwhelmed, and that can lead to depression. The first of the morning coffees kicks in as he reads emails from ER about potential discipline for the recently injured employee. Why bother with coffee? Why not just walk out of the office and find a real, and far less stressful, job someplace else? A wry smile crosses Aaron's face as he contemplates his ideal virtual life; owning a horse ranch in some remote corner of Wyoming. There would be no; injuries, people, difficult bosses; just feeding and shoveling with the horses he loves. Aaron's sneer broadens as he picks up his coat and heads to the door… The cell phone rings… "well crud, it's the Doc, 'guess I got to answer this one last call." "Breakfast and coffee in 15 minutes at the hotel?" Virtual escape sure sounds good, but real always seems to trump the dreams in the world in which he lives. Aaron heads out the door, but not toward Wyoming.*

The Doc relates some of his own background experiences that always are the precursor to some type of safety learning lesson. Aaron never knows for sure where the story will merge into the learning topic of the day, but he does know this process has never failed to deliver solution material for his own safety needs.

The Doc talks about working very happily and contentedly as the manufacturing engineering manager at the Fortune 20 company. Things were going well. The kaizen/continuous improvement teams of frontline employee technicians and supervisors were effectively splitting the many processes into unit operations, and then optimizing each step in each process, with some guidance from upper level management (the headquarters manufacturing engineering staff). Downtime had been dramatically reduced with each 0.1% of maintenance downtime reduction translating to about 0.4% increased operations uptime.

Delivering Safety Excellence: Engagement Culture at Every Level, First Edition. Michael M. Williamsen.
© 2021 John Wiley & Sons, Inc. Published 2021 by John Wiley & Sons, Inc.

And then a tragedy occurred at an East coast facility, when an employee broke a cardinal safety rule by doing her tasks in an automated operating crane process area. She got distracted and was crushed to death by a crane minutes before two senior vice presidents were to visit the plant. As these two men traveled back to the headquarters (HQ) location, they discussed the safety performance of their production culture. They were stunned as they remembered and recounted fatalities which occurred every three to five years and dismemberments which occurred every year. They had no safety department at headquarters, and no safety resources at any of the 40 facilities, and they had never really given deep thought about their Recordable Injury Frequency (RIF) in the range of 20. In their production-centric world, pay was a function of the big three and safety communication was nil. One of the tenets of their organization was to be number one or number two in all their product market segments. With a RIF of 20+ the company found themselves in the bottom third safety performance of all those in their industry worldwide. When upper management was told this fact, there was even more impetus to put an end to the poor safety performance. Additionally, this tragic trigger event brought them to the realization that the corporate culture they espoused, a fun food company, could not include killing or maiming its employees. Within a week, a surprised Doc was given the task of turning around safety at all the company facilities with no budgetary restrictions whatsoever.

After some research into industrial safety consultants, his small team settled on Dr. Petersen. "Dr. Dan" had a superb reputation in industrial safety, and the Doc was allowed to hire him as the consultant to help in this safety culture turnaround. The team originally met and worked with Dr. Petersen at this Fortune 20 consumer foods company where the Doc was in charge of turning around their poorly performing plant engineering and maintenance function at 40 manufacturing facilities spread throughout the United States. This engineering group was deep into a continuous improvement (CI) strategy which involved hourly and salaried technical personal at most of the facility locations. Performance improvements developed by the teams, and communications of what needed to be implemented for all facilities, were sailing smoothly throughout the field organizations.

As the Doc got to know Dr. Dan in this working relationship, it became quickly evident that Dan had only two tools: culture and accountabilities. These were mostly academic guesses as to what was needed, with no real industrial success cases to prove the academic theory. Nonetheless, the company's new, and hopefully ever improving, safety organization decided to use Dr. Petersen's two tools extensively, while also folding in the continuous improvement (CI), small team kaizen process. This CI culture was a part of the soul of the corporate organization culture for the mainline production and maintenance processes. Therefore, it must become an integral part of the new safety culture. Every safety process was broken down to a unit operation (fundamental safety process), and then each step of that fundamental safety process was optimized by the people who used it in the field. These small development teams, at a few developmental facilities, documented each process and its steps. They then communicated the solution process to other facilities which were charged with both implementing and improving (by using) what they had been given. Just as with the production and maintenance improvements, safety performance improvements were developed by the teams. The associated communications of needed safety improvements and their implementation at all facilities were sailing rather smoothly after about six months of intense work throughout the field organizations.

As time passed, and after Dr. Petersen retired, the Doc continued to struggle with the lack of an understandable, viable model which would represent the growth and development of an excellent safety culture. His experiences with past and present culture transformations in a variety of industries finally congealed into the six levels of a safety culture excellence model. The six level model in Figure 10.1 was purposefully linked to the 6 sigma quality excellence work of Dr. W. Edwards Deming and his compatriots. In the Deming model each sigma (standard deviation) represented an improvement in first pass product quality. One sigma being 68.5% quality (31.5% errors or defects), two sigma 98.7% first pass quality, and so on to 6 sigma, or three parts per million defects in the product produced. The higher the sigma standard deviation the better the first pass quality performance. The six levels of safety start with the basic reactive safety processes (as described below) and progress to level six where all levels of the organization were actively participating in their appropriate responsibility/accountability activities that help in developing and living a zero incident/zero at risk activities culture. After a few years of using this construct, the Doc was pleased at how well it seemed to work across various industries when an organization is committed to the relentless pursuit of zero at risk activities.

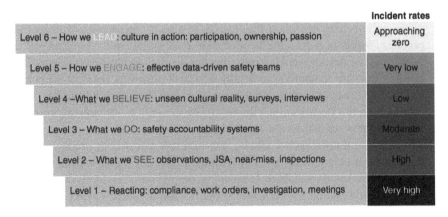

Figure 10.1 Six levels of safety.

- **Level 1:** *What conditions we react to.* This is foundational to OSHA regulations (regs). The regs are meant to protect workers from the dangers that exist around them. The regs provide detail about PPE (Personal Protective Equipment), LOTO (Lock Out Tag Out), working at heights, etc. All of the regs materials reference the dangers of working conditions, and what kind of protection is to be addressed in training and given to employees who work in and around these dangerous conditions. These kinds of conditions are truly the traps which can snare, maim, and kill working human beings. An organization must remove these traps, as well as train, protect, and enforce expectations that employees are at risk because of conditions. Additionally, the at risk employees must live a personal and team culture that follows the condition-centric rules which are designed to keep them from being injured. This goes back to one of H.W. Heinrich's principles: engineer, educate, and enforce. This philosophy has been a predominant paradigm driving safety for decades, yet has not been sufficient to eliminate injuries. Zero at risk activities require more than reacting to conditions. But reacting to and eliminating at risk conditions is an absolute foundational must in developing an excellent safety culture.

- **Level 2:** *What we see and react to.* One of the truly basic realities deals with working with your eyes open and your brain engaged. When this process happens employees should be able to react to dangers and avoid them. The many varieties of behavior based safety (BBS) focus on what dangers are evident in the workplace, and what the workers must do to safely react to the situations they encounter. Incident investigation, near miss processes, condition audits, Job Safety Analysis (JSA) work, observation events, and the like are all a part of reacting to what is seen. However, employees, supervisors, and managers still remain in a reactive mode, and unfortunately do not always react appropriately to existing dangers in the workplace. The culture continues to have at risk behaviors, and the associated injuries that accompany a reactive safety culture. Recognizing a hazard/risk frequently turns out not to be the problem. Workers seem able to recognize risks quite well. The problem is evaluating the risk and making the correct decision to modify one's behavior to mitigate that risk in addition to modifying at risk processes. Some of the better work in the area of reacting to the visible risks has been done by another good safety friend the Doc met along the way, Dave Fennell, a corporate manager of safety for Imperial Oil-Exxon Mobil Canada. Over a multi-year period, Dave researched and developed a 10-step personal risk assessment process. When this process is taught to and lived by frontline personnel, it delivers a significant improvement to safety culture. In so doing, employees at all levels address the 10 most significant hazard recognition and resolution issues which people face every day on and off the job. *This material is referenced in a following chapter* (Chapter 21).

- **Level 3: *What we know*.** Drs. Petersen and Bailey spent 10 years developing a diagnostic which measured safety culture reality in the workplace (see Appendix B). Their in-depth correlation studies provided 73 questions which map to 20 safety culture processes (categories) which, in turn, measure how well members of an organizations live what has been shown statistically to make a difference in injury rates. The 10-year study showed that organizations which have hourly, supervision and management who do the right activities described by the 73 questions, have fewer injuries. Each question is stated in an objective manner that delivers a "yes," or a "no," or an "I don't know" answer. In turn, all questions have a numeric rating as to the percent of the people at each level who have answered the question correctly. This quantitative number is used to deliver a profile of strengths and weaknesses of the safety culture across an organization's people (Figure 10.2), departments, sites, regions, countries, as well as a total organization roll up.

Frequently there is a desire for additional insightful learnings which can be achieved by combining the quantitative and qualitative analyses. In the qualitative mode, open-ended questions are asked at select slices of the organization. The people interviewed are those who should have important safety responsibilities, accountabilities, and experiences. Instead of asking "yes/no" questions, the interviews are comprised of "tell me about" questions which detail a person's and their group's safety-related activities and responsibilities. This additional information provides insights into the overall culture which really exists in the brains (between the ears) of those who make up, and practice, safety activities at the frontline and in the office environments.

- **Level 4: *What we do to improve*.** This step has the heavy lifting done by a cross section of people from throughout the organization. It is the process used by all levels of the organization, from executives through frontline hourly employees. Level 4 deals with what people across the organization do to resolve each stumbling block of their safety culture. After the level 4 diagnostics, it normally becomes very clear where the weak points exist in the safety culture. A leadership Safety Steering Team (SST) (see Chapter 13) that is comprised of hourly through executive employees then reviews and discusses the information and develops a strategic multi-year plan to attack the gaps that need filling. This same group recruits volunteers from across the ranks to work on, develop, resolve, and roll out the upstream process solutions for the issues they choose to address and fix. The SST meets monthly to monitor progress of the various CI solution teams. During this monthly day-long meeting they also engage in important developmental training in such topics as safety, culture, creative problem-solving, effective frontline communication, positive recognition, and the like. As the SST members complete their duties and members rotate off (and are replaced),

Score / Positive gap

Category	Emp.	Sup.	Mgr.	Emp. Sup.	Sup. Mgr.	Mgr. Mgr.
Recognition for performance	58.4	58.8	63.5	0.7	8.8	8.1
Discipline	60.9	77.4	66.1	27.2	8.6	14.6
Involvement of employees	72.3	80.1	76.4	10.8	5.7	4.6
Substance abuse	72.6	74.7	80.3	2.9	10.6	7.5
Goals of safety performance	72.9	80.6	79.4	10.6	9.0	1.5
Operating procedures	73.4	76.1	80.6	3.6	9.8	5.9
Supervisor training	74.9	85.0	84.6	13.5	13.0	0.4
Attitude towards safety	76.1	89.4	85.6	17.4	12.5	4.2
Inspections	76.4	90.5	85.0	18.5	11.2	6.1
Communication	78.5	88.3	85.4	12.5	8.8	3.3
Quality of supervision	79.5	89.4	88.1	12.5	10.8	1.5
Awareness programs	80.7	89.0	87.7	10.2	8.6	1.4
Safety climate	81.1	86.7	89.3	7.0	10.2	3.0
Support for safety	82.3	92.4	89.4	12.3	8.6	3.2
Employee training	83.4	91.0	90.2	9.0	8.1	0.9
New employees	84.1	92.4	88.2	9.9	4.9	4.5
Safety contacts	84.3	96.4	86.5	14.3	2.6	10.3
Management credibility	85.9	91.6	95.8	6.6	11.5	4.6
Hazard correction	86.1	92.3	95.2	7.2	10.5	3.1
Incident analysis	88.2	94.3	94.7	6.9	7.5	0.5
Combined score	77.6	85.8	84.6	10.7	9.1	4.5
Respondents	494	40	105			

Strong performance (≥ 90%) Needs improvement (75%–89%) Needs immediate attention (< 75%) | Needs attention (< 14% perception gap)

Figure 10.2 Results by category.

they return to their full-time jobs as knowledgeable, engaged, passionate safety experts. Level 4 is the strategic juncture at which "top down and bottom up, meet at the middle." Together they meet and defeat the significant safety problems that exist in all safety cultures.

- **Level 5: *How we engage and what we practice with respect to safety accountabilities.*** Accountability, "what you can count on me to do safely and correctly all the time that will reduce the possibility of injuries" is central to living a safety culture of what is correct, and what helps eliminate injuries. Dr. Petersen focused on this feature as an all important aspect of a proactive safety culture. Indeed, proactive safety duties are needed at every level of an organization. Of course, the frontline employees and supervision have significantly different accountabilities from those of the managerial ranks. Yet, just like what exists in a zero defects quality culture, all levels of the organization: include, practice, and sustain activities they do on a regular basis to eliminate the possibility of errors (incidents/injuries) occurring. This kind of proactive accountability is a dominant factor in excellent cultures, whether they be manufacturing, quality, military, or safety. These are the upstream activities which help deliver downstream results. In the 1980s and 1990s, when Dr. Petersen was espousing his safety accountability theory, there were very few examples of safety accountability he could identify in industry.
- **Level 6: What passionate safety leaders do.** The CI and SST team members learn hands on how to build an organization that lives a sustainable safety culture committed to excellence. In so doing, most become passionate safety leaders in their own work groups and within the total organization. This type of knowledgeable, passionate leadership from all levels of the organization drives a high-performance safety culture both on and off the job. They also quickly learn to apply the CI approach to the many other aspects within the organization, thus benefiting all the organization processes. A number of organizations have used the safety culture improvement models and CI techniques to improve other aspects of their company, as well.

10.1 Background for Culture Improvement

The Doc continued relating his story to Aaron about how he became deeply involved in developing effective, low incident rate safety cultures. He also found that without any safety department or plant safety people, the resource bench of talent was extremely weak for such a safety turnaround initiative.

In this Fortune 20 organization there was one person in ER (employee relations) who managed the worker's comp system, and he actually had a master's degree in safety. The one other resource at corporate was a former plant manager who traveled nationwide, addressing the many OSHA investigations which were the result of an unending number of serious injuries. Though he knew little about safety, this person did know a lot about negotiating OSHA fines.

As the team canvassed the company's 40 facilities they found only four plant managers who were willing to actively improve safety. From interviews with this group, the Doc began to develop best guesses as to the accountabilities at every level of the organization which would improve safety performance. The teams at these four "guinea pig facilities" tried them out and made adjustments on what would become the manufacturing organization's personal leading indicators. It rapidly became evident that an accountability culture was better, but was still not good enough. The clue as to what was missing came from the 6 sigma experiences of the many CI team improvements which had occurred in production and quality cultures. Their answer to fixing what was weak was to involve plant personnel in solving their own safety problems.

This "learn as you go approach" had the necessary executive backing for the safety culture to slowly begin to take shape in an effective manner. Within the first year a safety manual on the basics was in place. Accountabilities at all the developmental facility levels were functioning. Proven plant improvements to safety issues were published, funded, and implemented across the circuit, and injury rates were trending down. In year two of this safety culture excellence journey, seven more facilities signed on to stop the carnage.

The safety improvement team then came face to face with "the elephant in the room": cost and its effect on safety performance. Since safety was not a part of the performance reward system, there was no responsibility for assigning the cost of treating injuries at the facility level. All injury costs were assigned to a corporate account, which had no impact on the facility where the injury occurred. Contrary to this approach, if a facility spilled raw materials, or product, they were charged for these errors at the bottom line of their monthly cost runs, and thus their extreme focus on CI of all accounts the plant could control (with the exception of safety). Once again, upper management stepped in when prompted and the accounting system was changed to charge each facility for the actual cost of injuries of employees at its respective site. A corporate accounting group ran the workers comp (WC) numbers and began charging each medical injury at $1000 and each lost time at $25 000, with modifications at year end for the actuals of the settled and ongoing claims. As year three of the journey began, upper management once again came up to bat by dictating that all plant managers were responsible for implementing the new accountability and continuous improvement-based safety system. Additionally, they were required to provide a salaried, in plant safety resource. By the end of this year three journey the total organization was number two in their industry with respect to lowest Recordable Injury Frequency (RIF).

This intense three-year journey of developing and learning about safety, accountabilities, and culture development helped the Doc to understand a number of foundational requirements for improving a weak safety culture. These culture turnaround lessons seemed not to matter whether they were in safety, or any other underperforming facet of the culture. Basic culture truths are:

- Leaders, followers, and resisters exist in all cultures and change initiatives. You must get some leaders to move forward, and the leader of the particular culture improvement initiative must be one of them.
- What gets measured is what gets done, what gets rewarded is what gets done and done well.
- Recognition: gain sharing, trinkets and trash; in the long run more money or stuff just is not the issue. The solution is brief, genuine, one-on-one adult verbal appreciation for tasks and processes which are done well.

Recognition for doing things correctly seems to be a lost art. Over the years, the Doc assessed safety perception surveys for scores of organizations and tens of thousands of employees. As he tallied the results, recognition for performance of doing things right remained the lowest scoring safety management process. Interestingly, discipline (i.e. correcting people when they do something wrong) scored as the sixth lowest of the 20 safety management processes measured by the statistically validated survey Drs. Petersen and Bailey developed. So, whether employees do the job right or wrong, they are pretty much left alone to figure out what they ought to do. In turn, this approach leads to employees continuing the same weak "legacy" of safety processes and actions they have become comfortable with over the years.

Indeed, improving recognition skills is one of the low hanging fruits for an organization to improve the way its employees communicate important safety messages, which in turn help prevent injuries. During the development of the safety perception survey, there was an extensive effort to find a few questions which would reveal the real safety recognition culture in the workplace. The questions which were used in the survey as the benchmarks are:

- Is promotion to higher level jobs dependent on good safety performance?
- Is safe work behavior and attitude recognized by supervisors?
- Are safe workers picked to train new employees?
- Can frontline supervisors recognize employees for good safety performance?
- Is safe work behavior recognized by your company?

10.2 Human Interaction Realities

After reviewing the survey data from hourly, supervision and management employees, a continuous improvement team is launched. This team meets to develop safety processes and culture solutions and is made up of people from both frontline employees and management. One of the frequent Rapid Improvement

Workshops (RIW, another name for CI teams that take deep dives into solving complex issues) is to develop their own recognition system based on safety accountabilities, which are then practiced and positively reinforced every day across the organization.

A common thread is that many organizations have not trained their people well in the basics of human interaction. The symptom here is that personnel are not very effective in giving and receiving feedback on job performance, whether it is in safety, quality control, production, or off the job activities. A typical component of the team solutions then is to train all the personnel in giving and receiving feedback. Additionally, they must be trained about how to be effective in providing one-on-one recognition for doing a job well. The associated training and role-playing goes a long way to beginning a new culture of:

- Asking for permission to have the safety conversation.
- Getting a commitment to live safe behaviors.
- Following up in an adult manner.

In turn, this approach launches a coaching culture where hourly and salaried personnel try to observe and reward/recognize people doing the right things. All too often, safety pros, supervision, and management concentrate on what is wrong, with little or no positive feedback for the overwhelming number of times all is well with safety. The end result is that people know more about what the company culture does not want than what it does want.

An often used example is that of a coach. People are asked to think back to their coaching experiences, either as a player or as a coach. The effective coach watches what is going on and then intercedes where improvement is needed. This interaction is not punitive, but adult in nature. The player is shown what is correct, and then is expected to demonstrate this back to the coach until both are satisfied the basic skills are in place. The coach then continues to observe and give positive feedback as the player demonstrates the correct and improved skills. This simple coaching approach has many pieces to it:

- A one-on-one event focused on what is correct.
- A commitment from the student to do the task correctly.
- A consistent one-on-one follow-up on the skill in question.
- An adult approach to improving performance.
- A simple model that is used throughout to teach new skills related to safety:
  o Define the correct behaviors that eliminate unsafe acts and injuries.
  o Train all personnel in the behaviors which will improve both individuals and the overall safety culture.
  o Provide the necessary resources to improve safety performance.
  o Measure what is indeed being doing to correct and improve behaviors.

o Give recognition to individuals and their work groups as they accomplish these correct behaviors.

This approach is effective human interaction 101, but is seldom practiced in most safety cultures. Once the organization realizes the what, the how, the when, and the who, they almost always launch a successful initiative, which significantly improves not only their overall safety culture, but the other cultures (e.g. cost, quality, customer service, etc.) as well. A key concept to keep in mind is that visible upper management engagement and active middle management participation are necessary leadership resources for this engagement improvement process to deliver excellent results.

This process may seem to be a very detailed approach to move a current low personal contact culture to one of frequent, positive recognition for jobs done well. In fact, this is true. If you want something different, you will have to do something different. Organizations which have implemented an effective safety culture system have made positive recognition a part of their safety culture, and have involved all employees in the new process. In so doing, they have helped transform their safety culture to a healthier level that is performing well beyond their "old normal" process. The practical field experience is that instead of more money or stuff, most people would rather have a culture of realistic, concise, one-on-one verbal stimulation and recognition that is *timely, relevant, sincere, confirmed,* and *frequent*.

As organizations begin using the six level processes (Chapter 9) to improve their safety cultures they learn:

• Start small with a potential winner that you feel certain can be achieved. In so doing, your group can more easily build your team skills, experience, and morale. This is a sports team approach just like what is used by successful professional sports teams. In this way, you are able to begin a winner subculture which has the vision for the relentless pursuit of a culture of excellence.

• Do not begin with world hunger issues: "Don't try to boil the ocean, just deliver hot water in a cup sized portion." Where are the risks, which ones can we truly solve with the resources and talents we have available?

• While building the team, you are building resources to tackle additional and more difficult issues.

• As you succeed, publicize the successes and progress with leaders who will convince the followers to join in. Eventually, the resisters will have to change or leave, but do not ineffectively focus your scarce resources on chasing the "red herrings" of people or rare issues.

• While progressing through the six levels, should the teams utilize a series or a parallel approach? The traps that catch people are a definite priority, and yet an organization must get to engagement of the low-cost, active, people resources which normally exist in the workplace. The organization will have to work in a parallel fashion; it cannot wait for a series approach to get to level six. Remember, "The way to eat an elephant is one bite at a time."

- Frontline personnel bring practical experience while management contributes the resources needed to engage and deliver solutions. Significant barriers and gaps frequently exist between these two classic organizational levels. Fortunately, safety is a fundamental in most people's minds. This safety norm definitely assists in overcoming the interpersonal, multilevel barriers which exist. The end result being: working together on safety usually has much less resistance than launching a major change initiative in other operations related silo functions. It is *"top down and bottom up, meet at the middle."*
- How Many Teams? As many as possible, but… start small and do them well. Successful teams require:
  o Short-term (5–90 days) improvement teams
  o Effective facilitation
  o Effective leadership
  o Effective closure.

At a consumer products company in which the Doc had leadership responsibility, the need for improvement in every work cell was overwhelming. So, the Doc tried to do as much as possible, as quickly as possible, thus initiating 18 CI teams. This was an utter failure as neither he, nor any others, could effectively manage that many change/improvement teams at once. In the end, the Doc went to Tuesday and Thursday multiple two-hour CI team events. The teams then had the focus and resourcing needed to deliver solutions.

- A CI team culture approach is a unique tool to begin the organization on their journey to a culture of excellence. When the Doc became director of operations at a military ammunition manufacturing plant, there was a toxic relationship/culture between the union and management forces. The day he came on site there were more than 100 grievances on file, a safety severity rate greater than 140, quality and on time delivery were likewise in the tank. The Doc was quickly informed that a pool had already been formed on how long he would last (the previous six directors had lasted less than a total of six years), and the longest bet was only eight months out. Where to begin for this culture turnaround? The Doc chose engaging himself with hourly and supervision workforces. They worked together in solving the safety problems, all the while being honest with the union leadership. Neither the company nor the union could disagree with a focus on putting the workforce out of risk with respect to injuries. As credibility in these two areas was slowly and painfully built, operations performance and morale also climbed steadily. This difficult cultural shift also relied on the fact that the Doc and his team were honest in their communications and treated people as they should be treated in an organization that lives a sustainable excellence commitment culture in all that they do. Injury rates declined steadily while production and quality measures similarly improved.

Some of the keys to success were to deliver and reinforce an overall culture in which: honesty, principled treatment of people, engagement, and good communications were paramount daily aspects of all that was done, at all levels of the organization. These are basic fundamentals in delivering a new and improved world class culture.

Aaron feels somewhat relieved from the overwhelming morning tensions. At last he is getting practical, value added coaching on how to turn around a sick safety culture. The pieces of the puzzle are starting to fit into place, and the real world stories help him to move beyond the depression caused by his challenging situation. Others have had their significant struggles and been able to overcome them. Aaron realizes the escape option is not viable, no matter how appealing it is in moments of significant difficulty. The good news is that he can stand on the shoulders of those who have gone before him and learn from both their successes and failures.... It is still going to take a lot of work to do this turn around. Those stand on the shoulder stories all tell that there is no easy path to success in situations similar to what he finds himself. The "easy life" of a hermit shoveling after horses in Wyoming will have to wait.

11

Active Resistance

Once back in the company facilities, Aaron runs into many nay sayers, resisters, change haters, obstinate obstructers, theory X managers and a few other descriptors of the challenging people he deals with on a daily basis. Aaron has a slight smile as he muses over a few unrepeatable names for such people that the frontline workers use in his organization. All of these disturbing elements (or as some people call them, dirty elephants) seem to want their own personal piece of Aaron to tangle with. Wyoming where are you? Time to call the Doc and see if this chunk of culture turnaround is even remotely capable of being solved......

At dinner that evening, Aaron and the Doc discussed the inevitable attacks by people who resist change. The Doc commented about how for many years, with many different employers, he worked as someone who came into a company or organization which was in serious risk of going under. It seemed strange to him that almost without exception there was a significant resistance to making change/improvement in a culture where most all employees, at all levels, agreed the organization was terminally sick. Since the Doc was the change agent in charge, he saw lots of resistance from these organizations which had become comfortable with their weak status quo. After a few rounds of troubled companies, he came to a personal conclusion that most of these struggling cultures fit a 5-5-90 model. Five percent of the people would do whatever they could to help make the necessary improvements: the "developers." Five percent would do whatever they could to fight any changes: the "resisters/cavemen." Ninety percent would go in and out with the tide as organizational ebbs and flows occurred; "Just tell me what to do so I can do my work and collect my paycheck." This 90% group were like sheep following whichever shepherd was in charge. But in the beginning of a change effort the vocal, aggressive, negative groups always had the upper hand in a weak organization with poor leadership.

Delivering Safety Excellence: Engagement Culture at Every Level, First Edition. Michael M. Williamsen.
© 2021 John Wiley & Sons, Inc. Published 2021 by John Wiley & Sons, Inc.

The vocal and aggressive "resisters" seemed to never let up. The Doc commented about the discussion he had with Aaron before with the analogy of being an elephant on a tightrope performing in a circus (see Chapter 3). The gist of it was that there will always be some intense resistance to people who are leading a significant change initiative. The important thing for a successful improvement/change strategy is for the leadership to keep relentlessly moving forward and keep their focus on the goals of what is needed to be accomplished.

It turned out that the Doc's efforts and scarce resources needed to be focused on working with the "developers" who were able to participate in actively digging the organization out of the poor performance holes. Most of the Doc's time was needed to be spent leading the improvement efforts with those who would engage with him to do so. The rest of the time he just had to put up with the incoming negative efforts of the "resisters."

After about nine months of this approach, improvements took hold, performance improved, and the culture began to make a notable shift. Quietly and methodically over time the "resisters" no longer had the upper hand. The 90% follower group began pushing back when told to resist change and they slowly became advocates of improving the organization and its efforts. The Doc gave Aaron a quote that provided some guidance with leading a troubled organization from WOW (Worst Of the Worst) to BOB (Best Of the Best): "The needed sea change was visibly and palpably upon us when we engaged our people in focusing on and fixing what they knew to be wrong, while we let the flak throwers fade into the background."

The Doc added one more salient tidbit: at the beginning of every turnaround assignment it is wise to find out what your boundaries are and which barriers exist that require your attention while trying to improve a group which resists change. The explanation included an in-depth discussion and a few diagrams the Doc drew on a restaurant's paper napkin.

What are boundaries? A practical definition is "the lines or limits that are not to be crossed," such as *not passing a school bus when its red lights are flashing* and *one doughnut per week*. We all create boundaries, some less rigid than others, but they are meant to benefit and protect us without getting in the way of what we want and need to accomplish.

Nearly all of us struggle with effectively engaging others in getting our own desires to become a part of what these others do. There are lots of things we would like to accomplish, and in general, they require others' buy in. By setting up boundaries and rules, we are often faced with objections. It is natural for people to push back on our "reasonable limits/boundaries" in order to "get 'er done." We can expect resistance whenever our expectations are perceived to inhibit, or merely change, others' actions and wants. Yet some limits are needed. To help overcome this normal human behavior, it is necessary to have open, candid communication on boundaries, expectations, and common goals.

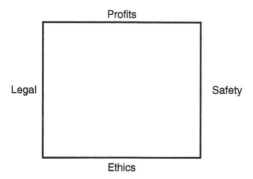

Figure 11.1 Four required business standards.

To illustrate this concept, the Doc was reminded of the boundaries that were once diagrammed by a former boss of his. The Doc knew his boss meant business because he asked for the door to be closed before taking a seat in the office. He sketched Figure 11.1 on a white board:

The boss carefully (and with some intensity) laid out what was bothering him about the Doc's work behavior. The four sides to the box represented the boundaries of the boss, inside of which were his comfort zones. He told the Doc that he must stay inside them. The four walls included safety, ethics, legality, and lost profit potential. He explained finite limits in each area and that he viewed all of them essential – to the company, to his career, and to the Doc's role as a manager. As the Doc looked and listened, he could understand the points of his boss, even though the Doc's boundaries, in some cases, differed from those of his boss. Both agreed that the Doc had never crossed any of his boundaries. (Had he done so, the discussion would have been about the end of his career for violating the rules of engagement.) Instead, the boss explained what was bothering him: the Doc was operating at (or beyond) the edge of his boss' comfort (buffer) zone (Figure 11.2).

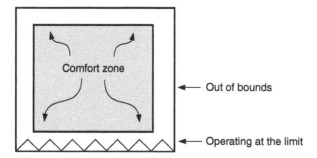

Figure 11.2 The comfort zone concept with respect to operating within business standards.

Although the Doc did not cross the lines, the boss' personal paranoia was that he had to have a buffer, or comfort zone, beyond which the Doc was operating. In truth, the Doc needed to give his boss a "mental comfort zone" that would not keep the boss awake at night, or provide worry about mishaps that *might* jeopardize the success of the business, have legal implications, or even cause embarrassment. The Doc's boss also emphasized that both the Doc and he needed to be well aware of all comfort zones and their boundaries.

From a rational management standpoint, buffer zones are a boss' friend (Figure 11.3). They allow some leeway for those subordinates who can operate too close to the edge. On the other hand, there are bosses who keep too tight of a leash because they are never fully comfortable with decisions their subordinates *might* make. These are the micromanagers of the world. Fortunately, the Doc's boss was not one of those.

The boss' point of view made sense, and in the Doc's mind there was much more to the diagram than what the boss initially drew. The Doc began to see all kinds of different boundary diagrams (Figure 11.4). Among the obvious ones were things like personal credit card debt or excessive speed. Another intriguing set of boundaries involves the raising of a teenager. It seems many children kept pushing the limits until they found out what "truth" really was. What they were doing was testing both the limits and the buffers of the parents. A "no response" from either parent set up a laissez faire family culture that was potentially dangerous to the children, the family, and others.

Aaron's comment was; "Ok, so what! What do the boundaries have to do with safety management?" In response the Doc added, "A prime example is the entire area of OSHA regulations." Here the boundary system appears inverted (Figure 11.5). Often the bureaucracy seems to care most about putting a check in the "regs" box. This is not the type of culture which leads to high performance. A healthy safety culture needs more than a "check in the box

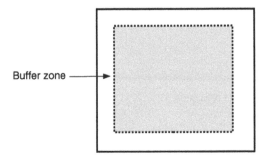

Figure 11.3 The buffer zone concept with respect to operating within business standards.

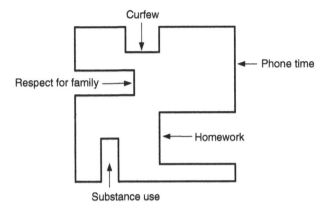

Figure 11.4 Amount of flexibility for a teenager with respect to family standards for behaviors.

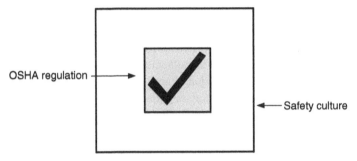

Figure 11.5 Safety regulations fit well inside safety culture standard limits.

mentality." Standards and regulations are only part of the equation. They are foundational and must be practiced, but an organization that lives a sustainable safety excellence commitment culture has to go way beyond just meeting regs and compliance work.

In conclusion, the Doc had never thought of his actions as out of the norm. For him, they were not. It was just that the Doc was more comfortable than his boss was when it came to operating on the edge. Rather than fight the proverbial City Hall (which never works), the Doc decided to give his boss a buffer in each of his sensitive areas. In the end, there was little difference in what the Doc was currently doing or how he was doing it. He was allowed to continue with a very similar strategy and tactics, but with a slightly longer time frame to achieve the goals both his boss and the organization needed. Had the Doc not changed to fit the boss' rigid boundary buffers (his needs), it is doubtful that the Doc could have kept his position much longer.

The Doc commented that dealing with the boundary limit paradigm has proven to be very beneficial as he has worked with people and organizations around the globe. Whether or not he or she realizes it, every boss has a mental boundary and buffer diagram that is not a perfect square. There are some areas in which a person has almost total free will, and others that are very well defined and limited. The Doc challenged Aaron to use a reasonable approach to identify the real limits as soon as possible. By doing so, one can extend his/her personality and passion into near unlimited performance. In turn, this understanding makes their life, job, and organizational experiences much more rewarding than living in a virtual, unexplored box of mediocrity with unknown boundaries.

During another drive home in deep thought, Aaron resolved:

- *To have a thicker skin*
- *To shake off the barbs sent his way (rather than let them send him into depression, or orbit)*
- *To meet with his boss to define the boundary limits and buffer zones*
- *To begin a dialogue with his boss about pushing the limits within those boundaries, so as to turn around the weak safety culture*
  As he pulled into the driveway, after another early morning and late night kind of day, Aaron wondered how long it would take for this transformation to occur. How long would he have to put on his thick skin and endure the slow march to delivering a culture which eliminated injuries in his company? And one other item; cost. Would upper management stay in the game funding this initiative when other projects were competing for scarce available funds?

12

Zero Injuries

After the long dinner discussion with the Doc, Aaron arrived home to be greeted by an unhappy wife. The long hours at work and short hours at home were definitely causing problems for all the family members and their two horses. She challenged him about him being married to his job and its effect on their relationship. Aaron had no trouble agreeing to that challenge. A late night glass of wine together and a non 5 am wakeup call provided some relief. Still, how long does it take to achieve a significant safety culture turnaround? The family reality with the loves of his life; his wife, the children and the horses, snapped Aaron into a better appreciation of his off the job necessities. Aaron did an instantaneous mental reset. The next day when he arrived at work at an unheard of late 8:30 am he called the Doc and posed the turnaround timing question.

The Doc's experience over the years was that there are two very different approaches to improving a safety culture:

o Living a principle which values the people who work for the organization.
o Reacting to a serious/fatal trigger incident that wakes up the organization's leadership.

The Doc's personal experiences with organizations that live principle and values has been the most pleasurable. When this approach is followed and led, it can deliver a serious, sustainable effort to eliminate all incidents which could lead to the harm of employees and the associated costs to the company.

This said, unlike at such organizations, most of the Doc's experiences in safety turnarounds have been the result of organizations reacting to inexcusable safety disasters. Safety has forever been a cost item in the day-to-day operations of safety reactive organizations. To such an entity it seems there is no perceived return on investment (ROI) for safety. And sadly ROI is what drives most operations investments for this group of companies. What kind of return can a company

Delivering Safety Excellence: Engagement Culture at Every Level, First Edition. Michael M. Williamsen.
© 2021 John Wiley & Sons, Inc. Published 2021 by John Wiley & Sons, Inc.

obtain from safety? In the business world reality where all organizations seem to always be competing for scarce funds, is an investment in safety better than one in capital equipment? Or what about the other perceived needs competing for funding and associated resources? Not likely, considering that the only safety strategy of any significance, for this kind of organization, has been one which deals with reacting to injuries and incidents. These costs are not realized or noticed until they become sunk costs, i.e. the monies have already been paid out by the organization before anything is even worked on.

The international operations world is undergoing significant changes. Organizations are ever more intensely competing on a global scale. The potential savings from advanced distribution technologies, quality improvements, and factory flow approaches are well known and practiced by all the viable corporate entities. Though these remain important factors in operations excellence, they are also often at the point of diminishing returns by all the major players. Labor in global and other large advanced industries is typically at a single digit percentage of total costs. However, those people who incur most of the industrial injuries are usually the ones at the front end of the manufacturing process continuum.

The people at this point in the total process are the ones producing the goods and services which fund the entire organization. The ever greater sophistication of equipment and technology, and the limited available educational training required to operate in these ever increasing technological environments, has put a new and different kind of squeeze on companies. The people able to competently perform the ever more sophisticated frontline jobs are ever harder to find, train, and retain. These are the same people who are protected in an excellent safety culture. Forward looking companies are recognizing the need to protect their living assets from industrial and off the job injuries. A viable bottom line truth is that these hard to find people have now become a part of a 5–10 year vision which details how an organization must protect their technology, quality, and personnel assets.

Back in the 1980s, the Doc was involved with a Fortune 20 company while it did an in-depth dive into safety costs. They found that each medical injury had an associated direct cost to the company of about $500. Lost time injuries (LTI) on average were much more expensive or around $25 000 direct costs each. An average fatality had slightly more than $1 million in direct costs. As they looked at OSHA data for their industry, it suggested a 5.1 : 1 multiplier of direct costs on average to cover indirect costs of the average LTI. This factor could quickly become a 10 multiplier with a dismemberment or fatality. Additionally, another one of the teams took a different approach in evaluating injury data. Their injury data-inspired model was that a near miss/close call was about 12 in. away from a medical injury, with a medical injury being only about 6 in. away from a fatality.

At a different company, the in-depth research of the injury cost runs easily justified a direct cost of more than $10 000 *on average* for each medical/recordable injury. As is common to companies beginning a significant safety initiative there is uncertainty as to how/where to start such a strategy. Figure 12.1 shows a

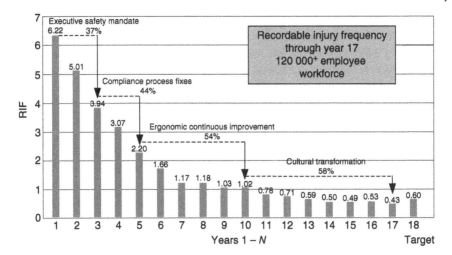

Figure 12.1 Safety journey.

plateau (pause in RIF results) for each phase of their strategy. This data from one of the Doc's customers was supplied by that organization's safety manager, Andrew (Andy) Schneider.

- At the beginning the executive level tried mandating (demanding) better safety (RIF) performance at every facility. This top-down approach reduced injury rates (reduced RIF rates) for two years before no more progress was occurring.
- Next they tried having the facilities improve each of the fundamental safety processes (training, injury investigation, and other compliance-based items). The organization continued to improve for two more years before this approach stalled.
- Ergonomic improvements were then instituted, this delivering two more years of reduced RIF before the next plateau was evident. After three years of little or no RIF reduction the large manufacturing entity bought a smaller company that specialized in helping organizations develop and implement safety culture excellence. As they implemented this cultural approach they were able to at last achieve the board of directors mandate of a company-wide (Global) RIF of 0.6 or less within 18 years. They even beat the time limit by five years once safety culture excellence engagement became the approach for all facilities.
- As detailed in Chapter 10, the six levels of safety performance require attention to reactive compliance-based issues, diagnostic evaluation of the true safety culture, and the implementation of an organization that **engages** all personnel levels in the relentless pursuit of a safety excellence culture. However, you do not have to wait 10 years to start implementing a strong safety culture initiative. Such a strong safety culture approach is available in year one of your organizations' efforts.

Both companies saved incredible amounts of money by eliminating injuries of their living people assets. A simple calculation showed that one injury at $10 000 direct and $51 000 indirect, on average saved $61 000+ bottom line impact. Another simple calculation showed how much additional product with its profit margin had to be sold to recover the cost of such an average injury. This impact for more than 65 000 fewer medical injuries to their living assets (people) is mind boggling to any upper level manager. Various data searches for costs never deliver precisely accurate results. However, though not exact, they provide good enough additional information to help inspire the actions needed to improve a weak safety culture. It is common knowledge that as time progresses the numbers, costs, and people realities related to injuries only escalate.

Can an organization only count cost savings? Indeed, cost is not all of the impact of an improved safety culture. What about the cultural impact of considering that the injuries could just as easily be to your own son or daughter if they worked as front liners in your organization?

A number of the organizations the Doc worked with began to note significant morale improvements in low incident rate safety cultures. In turn, this had them becoming the job hirer of choice for the needed frontline employees. Once the safety turnaround involved the frontline employees in solving the many safety issues they faced each day, these same people began applying peer pressure to the rest of their teams to live a sustainable culture of safety excellence. With this kind of culture the need for a safety cop function all but disappears.

The next step in the safety model surfaced from an employee, the 50-year corporate safety progression:

Is zero really possible?

- 45 years ago: Eliminate fatalities
- 25 years ago: Eliminate Lost Time Injuries
- 10 years ago: Globally excellent companies try to eliminate all medical injuries
- Today: A relentless pursuit of ZERO
  - Each Day
  - Each Week
  - Each Month
  - Forever

This simple progression model reinforces the necessity of becoming as close to a zero incident/zero at risk activities safety culture as possible. What is the target for this approach? It is making safety a core value for the organization. When safety is a core value, people believe that all injuries are preventable. When safety is a corporate value, companies strive for safety perfection. Companies that strive for perfection are committed to attaining zero incidents (zero at risk activities) and they actively work to manage business operations to achieve this goal. Using this approach is the essence of a culture of sustainable safety excellence.

It is rather simple. A company's commitment to everyone's well-being is reflected in its decisions about everything: from capital improvements and hiring the right personnel, to structuring for efficiencies, and supporting employees to get their work done safely and on time. A company that walks the talk goes beyond inscribing safety best practices into its mission statement. It must dig deeper and design every task of every job so that it can be performed with as little risk exposure as possible. This kind of company culture will continue to revisit, measure, and reward the living attributes of an excellent safety culture. It is as if their actions are saying "Our jobs and the company's success depend on the wellbeing of everyone involved."

The Doc recounted a personal discussion with Andy, the corporate safety manager of large global corporation. According to Andy their emphasis during the first 10 years of their safety journey (Figure 12.1) was on reducing risks and reinforcing safe behaviors. However, as they studied their plateaus they began to realize they were missing a key element of the process. They had over looked the key factor underpinning the entire process and making all elements of the programs work well together or poorly .... Company Culture. They found safety performance at some of the company facilities deteriorated when the leadership team changed. Some of their new company acquisitions did not improve at the same pace as others. Andy's group concluded that some of the gains were personality driven, rather than being truly embedded in the thoughts and actions of the company culture. These "personality factors" were essential for leaders as they interacted with and led their employees. The safety culture emphasized and led by the leadership made a significant difference in safety performance.

An important question is: whether the organization is confident in its process to review the performance of workers, managers, and supervisors. If such a process exists, is it on a par with what is expected from perfection processes in production, scheduling, cost containment, and customer–client retention? Is the organization's leadership talking about their legendary customer service in the same breath as their legendary leadership in safety excellence?

A paramount question for such an organization is: what actions are being taken to reward/recognize the desired safety performance and for continuous improvement? How strong and practiced is the safety performance feedback loop?

Does the company have the same level of expectations and processes in place to review supervisor performance around well-defined safety practices and duties? If so, great! They get it. Chances are, they are therefore rewarding the desired performance during performance reviews, bonuses, promotions, etc. Doing so establishes safety as a critical success factor in what the company values.

If this all sounds way too involved, difficult, and therefore inappropriate for protecting life and limb – or one more unnecessary nuisance to fuel the meritocracy, remember this: what gets measured is what gets done and what gets rewarded is what gets done first.

Indeed, as time progresses, the key performance indicators (KPI) of costs and people difficulties only escalate. So, the time to improve a safety culture is now, not living in denial and waiting until there is a disastrous trigger event.

As an example of this reactionary approach, there have been a number of articles and even books written on the many safety problems with underground mining in Africa, and more specifically in the Republic of South Africa (RSA). As the Doc's group was working in the RSA at a series of underground mine sites, they came away with the following conclusions:

The Republic of South Africa (RSA) has been struggling with reducing injuries in mines for a long time. Their desire for a "zero harm" (injury free) mine is a wonderful goal, but how can they go about achieving it? The many critical articles which have been written are frustrating to read when a safety consultant/expert knows the answer and no one answers the door when he knocks and offers to assist. The Doc and his group's on-site work in South African mines, and their meeting the individuals in the workface, has provided some insights into RSA safety culture, which also apply elsewhere. The Doc's experience around the world with long-established poor safety culture organizations provides practical experience in how to proceed in solving an organization's serious safety dilemmas.

An RSA article which the Doc read contains a synopsis which had many points he agreed with:

- Behavior based safety (BBS) is a sweet sounding panacea that does not necessarily work.
- There is a need to focus on mine behaviors with sound safety systems.
- You must have a basic level of coherence, trust, accountability, and a sense of fairness between management and labor.
- To a great degree the problems do not lie with the operators at the bottom, who are often accused of taking the risks and not complying with the rules. It is not an operator problem. It is an organizational problem, and above all, it is a senior management problem.
- "Every senior manager will tell you that he has bought into zero harm. In actual fact, many speak the language of zero harm, but don't really believe in it. The talk is there, but the walk is not. The driving force is production and when there is tension between production and safety, 90% of the time it's seems the production target is pursued, and because corporate level people are breathing down the neck of the mine manager, who is breathing down the neck of the shift boss, who is breathing down the neck of the supervisor, all the way down to operator level. And breathing down everyone's neck are the investors, who want production."

It is relatively easy to diagnose the problem, but how to effect the needed change? As the old saying goes, "The devil is in the details." The Doc's underground mine experience in South Africa told him that these organizations are open to change. In most organizations management cares about its workforce. They desire to see things improve. However, most executives around the world have no experience and no formal training on how to implement and manage safety. They do not understand the safety system and culture they are supposed to both manage and lead. As a result they have little to no understanding of the activities required. In a phrase; "They don't know what they don't know."

The answer to eradicating a culture of ongoing injuries is found in the development of excellence for: safety leadership, safety management systems, active appropriate safety accountabilities, and participation at all levels of the organization. A mission critical touch point for developing this safety culture excellence is executive education. Company executives must be taught foundational basics of *safety culture excellence*, which is very different from mere training on the regulations. In so doing, they will learn how to manage safety proactively through a system of accountabilities. Safety systems that work rely on involvement, empowerment, and participation of all employees; from the CEO to the frontline hourly worker. Management's role and responsibility for a "zero harm" safety culture requires:

- Management executives will need to stop managing inside the trap of staying focused on the downstream indicators of injuries they do not want to occur.
- Once they have shifted gears away from non value added downstream indicators, they must exhibit "visible, felt leadership," (a catch phrase that is commonly used and supposedly inspires all to work safely). In actuality this catch phrase is really just another way of expressing Dr. Petersen's criteria for safety excellence: "visibly committed senior management" and "active involved middle management." This kind of leadership revolves around activities which are engaged at all organizational levels. In turn, these effective safety culture activities come from a true understanding of safety culture performance.
  - This performance requires appropriate engagement in, and leadership of, the presence of safety activities which eliminate the possibility of injury.
  - Once these activities become the true safety culture there will be a relentless focus on the elimination of all workplace injuries.
  - At the start of such an initiative little of the above seems to exist and the team must commit to faith in the process as their organization proceeds.

The Doc's group did not honestly see that more regulations (regs) would be much of a help in the RSA. Yes, regs do give legal entities something to audit and give out fines on. The Doc just has not personally observed that this approach improved safety a whole lot at the frontline. While in RSA, the Doc's group had numerous opportunities to read some of the best safety procedures they had ever seen. Their FRPs (Fatal Risk Protocols) were most excellent. When the miners were quizzed underground, they had been well trained and knew the answers verbatim. When the Doc's group then went down into the underground mine, they watched the foremen and miners violate the FRPs with a total lack of conscience. This lack of care or focus on what is correct fed a workforce attitude that the Doc has observed in many struggling, autocratic labor-intensive cultures: "They pretend to pay us a little and we pretend to work a little." An engaged culture of correct leadership and labor workforce definitely does better than this.

The problem was a management focus on tonnage and little or no concern for the uneducated masses of impoverished African miners. These miners end up being required to do most anything asked of them in order for them to remain employed and earn enough money to support their starving families. It reminded the Doc's group somewhat of participating with safety work in Mexico, except the impoverished workers are not African miners, they are the Mexicans who live in the area of their employment. The solution in Mexico was to find facility managers who cared about people and then train them and their staffs in how to develop and live a zero incident/zero at risk safety culture of effective accountabilities. Once they understood and engaged, they quickly began working to achieve a zero at risk activity safety culture. The end result was their people living practical safety accountabilities at all levels of the organization.

The Doc's solution to the RSA mining fatalities (and with other very weak safety organizations) is three pronged:

1. Concentrate on the few organizations with a leadership that has the desire and willingness to actively put forth the efforts to get to zero at risk safety problems:
   - Train them how to do so.
   - Mentor and support them along in the process until they can stand alone.
   - As their downstream numbers get ever better, communicate their successes for all to see.
2. Continue heavy fines and shut downs for serious injuries and fatalities. This approach will force some of the others to seriously consider alternatives to their production first culture.
3. Give "Getting to Zero" webinars, conference presentations, and Safety Culture Excellence Workshops (SCEW) throughout the organization in an effort to educate and engage those who are interested in improving. In turn, this process will help push more organizations to follow the successful process.

This three-step process is pretty much the one successfully used by the Doc when he took over safety for a Fortune 20 corporation which had no safety culture, and was only focused on production throughput and cost control. It is not about a "one trick pony" regs approach, or about any other catch phrase. It is about a state of mind change and a culture shift that engages in safety value-added accountabilities and activities at all levels of the organization.

Aaron benefits from more explanation into the depths of achieving safety culture excellence. But will 10 years be required to achieve the desired outcome like the timeline shown on the chart? He's not sure his marriage can take it for that long. The Doc comments about Aaron and his wife making a commitment to spend time together off work, on a regular basis, doing stress reduction kinds of activities they both enjoy, and then sticking to that commitment 90% or more of the time.

Additionally, the first year is the hardest in implementing change. After culture improvement becomes a part of the organization's every day working culture, more and more of the personnel engage in fixing issues and delivering solutions. When this happens on a regular basis the leadership requirement for change has a reduced requirement on their personal time. This is because others are sharing the load. That is, except when an emergency occurs, but as more problems are solved and improved safety culture is sustained, fewer emergencies will occur. The first year or so is always rugged, always. Aaron will have to get through the first year.

13

How Long?

Aaron is back at work facing the personal struggles of one year with his nose to the grindstone for way too many hours, for way too many days. The slow progress currently provides no encouragement whatsoever. There has to be more to the process than lots of hours. He is wondering how to do the right things which will lead his organization to the right conclusion of a zero at risk safety incident culture. There remain critical missing parts to the puzzle. Aaron believes these critical process steps are certainly not intuitively obvious to the most casual of observers, or to him. This time he phones the Doc before heading out for the next breakfast teaching/training session.

The Doc begins with another story, this one about how his team worked with a large construction company that was building a power plant in a remote location of the country. They had begun practicing the process of achieving an excellent safety culture with well-defined accountabilities and frontline issue identification and resolution. During one of the random drug sampling events, a large percentage of the on-site iron workers showed disturbing positive drug test results. In-depth discussions brought forth the fact that skilled iron workers were hard to find in the surrounding area, limiting what could be done about the approximately 28% positive drug tests for this scarce skilled trade. They were already dangerously close to being behind schedule and could not afford to be short of this trade's work.

The expedient answer was to live with the problem and move forward. However, when principles of worker safety, ethical and legal standards, and living attributes of an excellent safety culture were brought to bear, it was decided to terminate ('fire'/'let go') the problem tradesmen and face the consequences of living with principle-centered leadership. For about three months, installation efforts lagged. Then a wholly unexpected dynamic began to emerge as iron workers from other jobs in the region began answering job placement advertisements. They had heard this was a construction site where leadership was exceedingly safety conscious and was taking serious initiatives to make and keep it that way. Shortly thereafter, the

Delivering Safety Excellence: Engagement Culture at Every Level, First Edition. Michael M. Williamsen.
© 2021 John Wiley & Sons, Inc. Published 2021 by John Wiley & Sons, Inc.

safer workforce, which was engaged in delivering solutions to all kinds of safety and non-safety issues, was ahead of schedule. Ultimately, they ended up delivering completion of the project, not only ahead of schedule, but with one of the best safety records in the region. As a consequence, there was a significant monetary performance bonus for all to share, besides the satisfaction of everyone going home every day without safety incidents

Make no mistake that taking people out of the workplace will likely affect production in the short run. Because of this potential problem, a very real struggle typically emerges concerning the staffing of continuous improvement (CI) teams. Production wants their people engaged in *meaningful* work that delivers profits to the bottom line and not spending hours in meetings which deliver no product. The Doc and his teams have engaged in this arm wrestling struggle countless times.

As an example from a different view point, hunger pangs typically lead down one of two paths: get it now and pay the price in your gut and wallet, or apply the proper ingredients and time to create something you will enjoy somewhat soon and again in the future. This second approach is reminiscent of the two options to making popcorn. To make popcorn you can:

- Pour in the kernels.
- Set the temperature according to a well-defined and tested process and let the kernels pop in their own timing and enjoy a well proven result.
- Jack up the temperature forcing the kernels to pop according to your timeline and quickly deliver substandard popcorn by hurrying the process with resulting burned popcorn which nobody truly likes. However, we do deliver to the demand of "I want my popcorn and I want it now." .... Even if very few of us really like the quality of this quick focus end result.

That said.... the successful approach to improving weak safety processes has a number of well thought-out steps. Each of these steps takes both concentrated time and effort to move the safety turnaround process to success. The list given below provides an introduction to the in-depth materials that follow in this and subsequent chapters.

1. First, perform a diagnostic to determine needs to be worked on according to what employees believe from throughout the organization at all levels (Chapters 7 and 10).
2. Next, a Safety Steering Team (SST) is formed which has membership from throughout the levels of the organization. This group regularly meets to determine a long-term improvement strategy with agreed upon priorities (Chapter 17).
3. Then Rapid Improvement Workshops (RIWs) are formed which are also made up of interested volunteers from each level of the organization. These RIW teams tackle one to three initiatives each year. They spend about 90 days

deciding what to do, how to do it, and then fleshing out the details of the solutions (Chapter 18).

4. Next the proposed solution is field tried (piloted), improved, and proven at an existing actual work site. This step takes about 60 days to get hardened, workable successes.
5. After the solutions are proven other sites implement the solutions over time.
6. All the while, the SST monitors the progress of the individual RIW efforts as well as the pilot trials. They also decide the launch particulars of the next RIW in line and report the results of this well-managed improvement process to intermediate and senior leadership.

It is no surprise that many organizations are under pressure to immediately solve their weak safety culture issues. They must stop the associated financial, personal, and business loss bleeding that is resulting from their poor safety performance over time. However, there is a decision to make after the diagnostics are done:

- Following this proven safety culture improvement process with its time requirements, or
- Diving in with a management team and immediately implementing their best unproven, expedient guess as to what they believe will work (program of the month).

Realistically, the expedient approach seldom works and the classic definition of insanity usually rears its ugly head as another flavor of the month safety band aide fails to stop any of the bleeding.

Over the years, the process the Doc uses, and helps others to implement, is a time-tested, value-added, proven approach to spending scarce resources. Difficult issues which have degraded over the years take time and effort to address and deliver a solid, sustainable resolution. These weak safety processes may have been underperforming for a long time. If you take a look at the chart back in Chapter 12, you will note that significant resources and focus were required by a major organization over 13 years to reduce a RIF of 6+ down to 0.5. There is no sustainable quick fix to save 65 000+ employees from serious injuries. It takes time, effort, engagement, passion, perseverance, and the utilization of excellent safety improvement processes to accomplish this kind of a significant result.

Figure 13.1 shows progress toward a strong safety culture by a number of different company facilities.

Aaron noticed that the timing to get better was much less than what he observed on the other charts. When questioned, the Doc commented it was the result of using the CI tools at every stage in the improvement process. The Doc then stepped back and said it was time to introduce Aaron to some of the simple "tools" that people in an engaged safety culture use to develop excellence

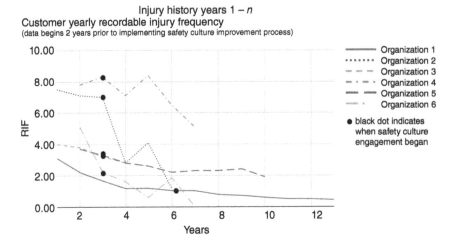

Figure 13.1 Recordable injury frequency chart.

in all they do. He explained this was an introduction (intro) to be used as food for thought while Aaron and his committed leadership began their journey to world-class performance. As they moved forward these same tools would be explained in greater depth when they were fit in with the other models and techniques for delivering safety culture excellence.

One of the first critical processes is getting viable people resources to do the necessary leadership and continuous improvement team work. How to effectively approach the correct decision person to dedicate work time (or as the frontline employees typically say, "get work release time") is a function of who that decision person is and what their style is. After deciding on an approach, or two, you will likely have to "keep coming back to the well" until you get the right people, on the right project which will allow you and your team to go forward. Beware of the supervision and decision maker trying to have you staff your team with their losers or firebrands. You need volunteers, and likely a few "voluntolds," who want to help and will put forth the efforts to solve the issue, not people who merely attend meetings or cause chaos. You desperately need the 5–10% of the people who will help you drive to success. They are the solution developers. The rest of the organization will become the implementers of the processes your team is going to develop. With this in mind:

- Pick a topic you can deliver with high certainty.
- Pursue only one topic, but make sure it has support from the decision maker.
- Run no Rapid Improvement Workshops (RIWs) during crunch times which occur typically at the end of a quarter, or year, or during a storm, or other temporary crises.

Figure 13.2 Purpose Outcomes Process.

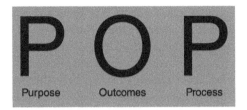

- Make and keep a list of what needs to be fixed while going through these high-stress, low-CI team work times.
- Develop a meeting schedule which can be supported by the decision maker. It is usually best to stay away from Mondays and Fridays. Doing the team meetings on a Tuesday or Thursday, for a maximum of two hours, has often been found to be close to the optimum choice.
- Inform, and frequently communicate with, the operations supervision about the schedule, and hold them responsible for the necessary backfills. Expect resistance, especially on meeting number one. Stay in the loop with the decision maker so neither of you is blindsided by the last minute refusal for work release time.

As your team moves forward, force communications with team members, with the decision maker, and with the supervision. Use, and communicate appropriately, the "POP" (Purpose, Outcomes, Process) statement (Figure 13.2) and an Action Item Matrix (AIM) (Figure 13.4) to show the issues, the progress toward solutions, and what remains to be accomplished.

13.1 POP Statement

Timing
POP is an acronym for: Purpose, Outcomes, Process. Often this initial work on the POP statement can be completed within an hour at the end of meeting number 1, following the delivery of the SST training. This process is very similar to the POP statement development for a RIW/CI Team, except that it is specific to the work of the Steering Team. The team's work begins by clearly stating its mission. The tool to accomplish this is the POP statement.

Purpose
Facilitate a discussion around the purpose of the SST. The SST purpose should be an overarching statement describing the team's mission to oversee the transformation journey from good to great. Sample purpose statements are included in the sample Charter Document shown in the Appendix C. Be careful not to share those samples with the team too soon. It is a healthy exercise to have the team go through the creation of a team purpose.

One way to facilitate this discussion is to have each team member write on a 3×5 card what they think the purpose of this team should be. Next, write those statements on a flip chart or in the Excel spreadsheet so everyone can see them. This process will generate good discussion around the words team members have used, and it will help the team brainstorm toward a compelling mission statement. Another approach has the facilitator visually going around the table asking each team member for input and discussing their input with the rest of the team. This kind of interactivity helps build team work performance, which is critical to success. Depending on your team members' proficiency with this kind of interactive continuous improvement work, it may be a good idea to share some basics of the process like: *share openly, no criticisms of thoughts/ideas expressed, build on one another's thoughts/ideas,* etc.

Outcomes:

There are two basic types of SST outcomes: long-term sustainable outcomes and short-term outcomes. Long-term outcomes refer to the strategic ways the organization will be impacted overall by the transformation effort. Short-term outcomes deal with near term things the SST and RIWs must complete to be successful.

Process:

What is the process by which this team will operate? Things like when, where, and how meetings will occur; ground rules for meetings; how many CI teams per year will the team commission; how will the team make decisions; how will they hold one another accountable for the work, etc.

In the Figure 13.3 example, the "A" items have highest priority and so on down the alphabet. Many teams choose to just number the POP action items and have the team members flesh out what needs to be done for each Outcome bucket.

Team Update	October 11		
Purpose	Develop a comprehensive inspection system that identifies all hazards to help us achieve zero injuries		
	Rank		**Comments**
Outcomes	C	A training program set-up by safety for the new inspection system	To be determined
	A	Gather existing checklists	Complete – August 8
	A	Team reviews existing checklists and decides to leave them be	Complete
	C	Develop the system present to management for critique and approval	Ongoing
	B	Develop inpection accountabilities for all levels (in Policy Statement)	Due – October 15
Process	Use a hybrid kaizen event to develop the improved process and the 4 steps to accountability to build a robust pre-planning and operating procedures process		

Figure 13.3 POP statement: CI inspection team.

Inspection CI team						
	Action Item	**WHO**	**Target date**	**X**	**Comments**	
1	Next meeting 8/1 at 1pm. Reserve a room-Mead 1	Norm	7/27	X	Done	
2	Type notes and prioritize outcomes	Frank	Before 8/1	X	Notes typed, Items prioritized 8/1	
3	Review flip chart (figure out hard to read words)	Team	8/1	X	Done – 8/8	
4	Team Leader/Facilitator	TBD on 8/1	8/1	X	Norm – Team Leader, Frank Secretary	
5	Finish off action item matrix		On going			
6	Develop physical inspection for each dept					
7	Develop physical inspection for each dept					
8	Get other assistance for inspection					
9	Develop our process for doing this inspection/flow chart	Steve and Ross - draft flow chart	8/15	X	Done-team reviewed "fault-tree"	
10	Bring up to 3 inspection sheets used in your department area	All Team Members	8/8	X	Done	
11	Review inspection sheets - ok for dissimilarities, redundancy, good info to future use etc?	Frank	8/15	X	Inspection sheets form different dept are "all over the map"	

Figure 13.4   Action item matrix.

13.2   Action Item Matrix (AIM)

The AIM (Figure 13.4) is basically a "to do list" for the team. It captures **what** needs to be done by **when** and by **whom**. It is used at every meeting to serve as the mechanism for accountability and follow-up. The team leader will review the AIM at the beginning of each SST meeting to follow up on commitments made, and also at the end of each meeting to ensure new commitments and actions made during the meeting have been properly captured. The person who is recording the notes can also be the keeper of the AIM, but some leaders prefer to do this themselves. Either way is fine, as long as the AIM is maintained and used properly at each Steering Team meeting.

If one of those unforeseen crisis events occurs, be prepared to be flexible, while at the same time reinforcing the commitment to the next team meeting. The Doc's experience is that by meeting every other week the team members have time to work on their action items. If you meet weekly on these deep dive issues there is not enough time to dig through all the materials necessary to reach closure. Additionally, the production group struggles more with weekly work release time than with biweekly events. If you meet once a month there is not enough pressure on individuals and the team to address and close the action items in a timely manner.

As an extreme example (Figure 13.5) of this kind of deep dive, a team Doc had worked with in the past spent three to four months to develop a solution on getting to a Workers Compensation (WC) solution using a flow chart graphic. All this

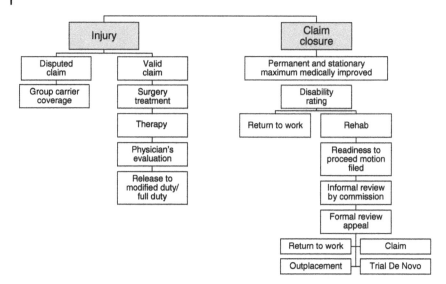

Figure 13.5   Workers' compensation carrier claim processing procedure.

was done in an effort (purpose) to close out the open cases, which had a history of dragging on for years. This entire process required input from a number of departments: legal, employee relations, the workers compensation carrier and facility leadership. This proposal was followed by two to three months for two corporations involved to test/pilot the solution, then multi-months for both organizations to roll it out.

13.3   Workers' Compensation Carrier Claim Processing Procedure

As time goes on, developing successful solutions to multiple troubling issues with this RIW approach will have less resistance from leadership, supervision, and frontline employee participation when all can visualize the entire process including an end point.

It will take multiple successful team solutions to change the operations culture, which typically is very much entrenched and noticeably difficult to change. Highly resistant to change organizations never seem to get better; they just keep the barriers in place and continue the Einstein definition of insanity. To overcome this resistance you must typically begin, provide strong leadership, and stay the course for 9–12 months before you can see the culture begin to shift for the better.

This brings to mind a challenging manufacturer of explosive materials where the Doc was brought in to do a turnaround. The union had a small group of hecklers who did an excellent job of researching all the mistakes that were made and throwing them back in the Doc's face at the monthly all-hands meetings. Each month, because of their proficiency in giving him grief, this all-hands meeting was truly painful for the Doc to direct and lead. He had to communicate with multiple entities and had to stay the improvement course, a process required of all change leaders. The Doc did pay attention to their legitimate flak, and indeed even did use the issues they brought up for CI team focus.

After about nine months of taking the grief, which always seems to come with being the change agent leader, there was a significant shift. At the monthly all-hands meeting end there was always an open-ended question and answer (Q&A) section of the Doc's time in front of the employees. Immediately, one of the hecklers (Gloria) started to stand up. To the Doc's surprise, the man next to Gloria put his hand on her shoulder and told her group to "Give him some slack; things are finally getting better after decades of misery. So give the guy a chance to keep the improvements going." This first breath of fresh air was phenomenal to the Doc and his teams of hourly and salaried volunteers. It re-energized all of them. They had finally gone from "complaint equals grievance" or "complaint equals BMW" (Bellyache Moan and Whine) to "complaint equals goal." Their goal of an organization that lives a sustainable excellence commitment culture in all that they do had begun spinning up, and stopped spinning down in the death spiral which had been so difficult to overcome. At the end of the first 12 months, not only were production numbers up, but the injury severity rate had been reduced to 9 from the 142 level. During this one year, the teams (Figure 13.6), not the Doc, had completed more than 600 action items. The next year they did more than 700 action items as production, quality, injury rates, grievances, and morale all made quantum shifts of improvement.

As a wrap-up on safety culture using the engagement of an organization's employees the Doc showed an easy to remember model: "How to start a fire."

Figure 13.7 is the classic fire triangle; let us use it for the ease of the analogy without getting into technical details. To start a fire you will need at least three elements: oxygen, fuel, and heat.

The other half of the analogy (Figure 13.8) deals with starting and maintaining a culture that is on fire for a culture of safety excellence.

- The counterpart for *oxygen* is the involvement of people in your organization who want to assist in becoming an excellent zero at risk safety culture.
- The *fuel* for these people who want to get much better is the risks that exist in the organization. The safety perception survey diagnostic, individual interviews,

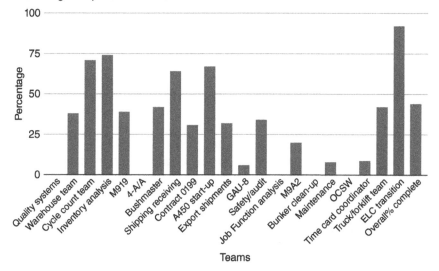

Figure 13.6 AIM team tracking.

Figure 13.7 The classic requirements for having a fire.

Figure 13.8 Requirements for having a culture of operational excellence.

safety work orders, Action Item Matrices, and the like are all risks that need to be addressed and resolved by your involved volunteers.

- The practical, leading indicator ***accountabilities*** your teams of safety process developers deliver as outcomes of the Continuous Excellence Performance (CEP see Chapter 16) process become the sustainable **heat** which helps to deliver an excellent safety culture that just does not accept anything other than excellence.

The practical experience with "on fire safety cultures" that engage their people in the relentless pursuit of safety excellence has these three ongoing elements firing on all cylinders every day.

> *Aaron once again benefited from a discussion on the how and flow of leading cultural improvement initiatives. He wrote down the rules of engagement and copied the various charts like POP and AIM, which will help his improvement teams stay on target and deliver timely solutions to the many, many holes in the company's filter barriers. So far all this material was coming together and that will help to improve Aaron's weak safety culture. But what is the ultimate goal and are there still more steps to take?*

14

World-Class Safety

Aaron goes out into the field and observes, as well as interacts with, a number of work crews. Morale is low. The crews know that the big boss and ER are back in the same old game of punishing those who are unfortunate enough to be injured on the job. One of the old salts Aaron goes fishing with on occasion stops him and says "Discipline, punishment, kangaroo courts. Why even bother with trying to improve safety? The same frustrating game of good cop-bad cop has been going on forever. Does management really believe 'punitive inspiration' is a viable leadership process? If you have a vision, what is it, and how are YOU going to achieve it?"

Aaron launches into teamwork, continuous improvement engagement, world class safety and . . . . .. stops in his tracks. Aaron has talked about it with the Doc, but Aaron doesn't really know the "how" or the end game "what" in a way he can credibly share with the frontline employees to effectively communicate that everyone must work together to protect each other. "Ok Doc, let's see you explain this one!"

The Doc's personal story describes a time when he was working with an underground mining customer in a sub-Saharan country. This organization had a long history of serious injuries and fatalities. There never seemed to be an opportunity, or a hope, for the impoverished, uneducated miners who, day-in and day-out, were tasked with the manual hard labor jobs common to many underdeveloped countries. Upper management seldom went underground to observe working conditions first hand. They felt that the miners were just numbers who could be easily replaced when injuries occurred. The Doc remembered listening to one of their mine bosses complain to him that in the past, they did not have a significant injury rate when they could "beat the miners." His explanation was that in the apartheid days, supervision carried around rubber hoses to get the attention of those who needed correction. Since the government had changed and outlawed this practice, they were struggling with how to make an effective safety

Delivering Safety Excellence: Engagement Culture at Every Level, First Edition. Michael M. Williamsen.
© 2021 John Wiley & Sons, Inc. Published 2021 by John Wiley & Sons, Inc.

culture change. In other countries and in other industries, it seems that when management views frontline labor as easily replaceable numbers, rather than real people, this kind of callous culture is replete with tragic safety incidents.

A culture that focuses on what the employee did wrong merely emphasizes fault-finding. Such a group is seldom, if ever, in the fact finding mode that leads to identifying and solving underlying issues which lead to injuries. With faultfinding there are no lessons learned to prevent the recurrence of a related injury. There is no focus on how management resources could have been used to help prevent the incident/injury. There are only the facts about when the next similar injury occurs that "There is no education from the second kick of a mule."

The safety pro who focuses on punishing (disciplining) others for rules/regs infractions is missing a whole new world that can be so much better. This new world of performance is based on a personal discipline of living a culture of excellence in all that is done, every day. In this kind of high-performance safety culture, a better definition of effective discipline is: "Training which corrects, molds, strengthens or perfects." This type of culture is a world that focuses not on punishment, but on a personal discipline which lives and executes excellence in activities as the acceptable goal of the work life. The safety pro, and the leadership staff, need to put an end to Theory X management and move on to much more effective Theory Y and Theory Z approaches.[1] In this new world, unless the infraction is a flagrant abuse, discipline for correction becomes a culture of coaching on how to do the job correctly, and an accountability for follow-up to see that the lessons have been learned and applied. In turn, this approach leads to a culture of a personal, not punitive, discipline in which employees at all levels of the organization are always endeavoring to do the job correctly and safely.

Moving beyond the punishment issue, it is unfortunate that people involved with safety have many times experienced the result of personal tragedies to them-selves or others. These trigger events are often like the tragedy which Aaron's organization suddenly became involved in. Will Aaron's organization wake up and decide to put forth the efforts to end these inexcusable injury incidents, whether or not, the injury was small or tragic? Aaron and his old salt friend know all too well this kind of commitment does not always occur.

All too often, such an initiative to improve safety falls short because the objective of the organization is to improve safety based on injury rates. The Doc's experi-ence with effective safety initiatives is that their objective must be to improve the safety *culture* into world-class excellence, rather than to focus on RIF. The RIF numbers game always exists to some extent. However, a single focus on reacting to what you do not want to occur is much less effective than a long-term focus on reinforcing what you do want to occur. This approach requires focusing on the upstream activities that deliver downstream results. The RIW/CI team activities which resolve issues such as those commonly found with equipment, process, and training. In this way the necessary changes are developed by the people who live in the struggling culture. This solution approach is much more effective than chasing safety RIF numbers. This is especially true when the teams are able to deliver value-added personal accountabilities, which, in turn, become the active leading safety indicators that help prevent incidents.

After a tragic event, the serious organizations realize they can no longer tolerate the insanity of status quo. Most successful organizations want to be number1 or number 2 in their markets, to be the leaders. When they do an in-depth data search, the typical result is that they are in the bottom ½ or worse of global safety performance for their industry. This information should be another wake-up call from the culture of safety insanity that seems to follow a lagging indicator reactive model. True excellence requires much more than regulations and compliance focused safety culture norms.

What are your safety culture realities that relate to what is truly the between the ears mental evaluation of safety as seen by your employees? Organizations which are leaders in the three frequent production characteristics (cost, quality, and customer service) have the capability of being safety leaders as well. They know how to engage their people in improving the cost, quality, and customer service pillars of excellence. However, all too often, safety has been mired in a culture which views safety only as a cost center. In this scenario safety is perceived as not core, but a "bolt on" that is shoved down the throats of productive cultures. Not surprisingly, the typical production culture reacts negatively to bothersome, low knowledge "leadership" which keeps tossing on, and painfully enforcing, more sets of ineffective rules. Those organizations that can overcome these kinds of bellyaching attitudes change their focus from a BMW culture to one of active involvement in fixing what needs to be improved. Instead of an ineffective focus on complaining, excusing, or punishing, they relentlessly do what is required to "find it-fix it-and move on." Responsible organizations do not live with poor performance in anything.

Set your goal to be number 1 or number 2 in safety and put forth the long-term effort and initiative for achieving your goal. Research what the global leaders have done and steal ideas shamelessly; you do not have to reinvent the wheel. However, your organization and its people may need to invest significant effort to overcome the multiple decades of entrenched poor safety culture created over many years. Likewise, it will take multiple years to develop a robust culture of excellence.

An indicator of the difference between a reactive and proactive safety culture is shown by the chart in Figure 14.1 developed by one of the Doc's fellow consultants, Todd Britten.

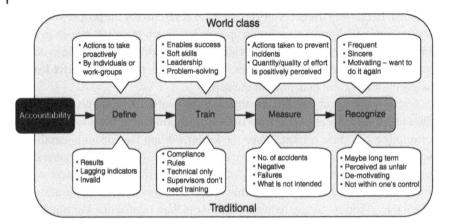

Figure 14.1 Accountability: World-class vs. traditional.

There is a huge difference between the traditional and world-class approaches to safety accountabilities. From the chart, the traditional approach stays in the reactive mode focused on:

- **Lagging indicators** as a definition of safety excellence. The world-class approach has well-defined proactive accountabilities which reduce the probability of incidents occurring for both individuals and work groups.
- **Training** in the traditional safety culture deals with compliance issues, which in turn, are predominantly a reactive mode to conditions in the workplace. The world-class accountability model trains frontline hourly and supervision in compliance issues as well as soft skills, leadership, and problem-solving.
- **Measuring** in the traditional model deals with reacting to incidents/failures that occurred in the past. The world-class approach does include reacting to events, but also focuses on the quality and quantity of efforts taken to prevent all incidents.
- **Recognizing** and reacting to what is not desired to occur, incidents and injuries. These are often not within the control of those being judged for safety performance. The world-class model maintains a focus on reinforcing what individuals and work groups are doing to prevent inexcusable injuries (aka all incidents). This is accomplished through personal ownership and implementation of safety-related actions.

What is world-class safety? When the Doc worked with Dr. Petersen he asked him this same question. The consumer foods company the Doc worked for decided they would do what is required to be world class in safety. In the 1980s, Dr. Dan's answer was along the lines of: "The best have a RIF of between 1.0 and 1.2 with a lost time frequency 10% of that or 0.1." When asked about the origin of those numbers, Dr. Petersen's answer was that they were the best 10% of his customers. This was a good interim target for the consumer foods organization with a RIF of 20+. However, as they got better and better and looked at how their teams made all the necessary improvements, it was not by focusing on chasing reactive RIF numbers. They achieved safety excellence by: "Living a continuous improvement culture which engages our people in the relentless pursuit of zero, every day in every way." This then has become the Doc's definition of world-class safety.

> *Aaron likes the concept, but is still unsure of how to proceed. With all the theory X leadership he has experienced, especially since he became the safety manager, Aaron is more than just a little gun shy. He is willing to move forward, but wonders about the types of hidden traps that can potentially lead to a disastrous conclusion, rather than to a world class continuous improvement engagement safety culture.*

Note

1 From a Wikipedia source: Theories X, Y, and Z deal with human motivation built on Douglas McGregor's Theory X and Theory Y. Theories X, Y, and various versions of Z have been used in human resource management, organizational behavior, organizational communication, and organizational development. McGregor's Theory X states that workers inherently dislike and avoid work and must be driven to it, in contrast to Theory Y which states that work is natural and can be a source of satisfaction when aimed at higher order human psychological needs.

   One Theory Z is William Ouchi's so-called "Japanese management" style, which was explained in his book *Theory Z: How American Business Can Meet the Japanese Challenge* (1981) responding to the Asian economic boom of the 1980s. For Ouchi, Theory Z focused on increasing employee loyalty to the company by providing a job for life with a strong focus on the well-being of the employee, both on and off the job. According to Ouchi, Theory Z management tends to promote stable employment, high productivity, and high employee morale and satisfaction.

15

Watch Out

Aaron realizes this is his last day with the Doc, but there is so much more to learn. The reinforcing support has been uplifting beyond anything he expected. But what will happen next week when Aaron doesn't have a fellow swamp rat to lean on? The cultural difficulties are still rampant in the organization where he works. Last night, Aaron had trouble sleeping as his mind conjured up all kinds of potential pitfalls that were "certain to happen" the moment the Doc left. He has learned that hiding under the desk does not solve the problems. In his gut Aaron knows, sooner or later, he is going to have to lead the charge for change, no matter how stressful it may be. Aaron knows "Sooner" certainly looks more like what is about to happen. Maybe the Doc can provide some insights on overcoming the demons that are hiding in the closet? Aaron wonders how to begin with this improvement process which all departments want, and they need help to achieve?

15.1 Setting Priorities

The launch point of a safety culture turnaround contains significant risks. This time the Doc's story was about him working as manufacturing engineering manager for a consumer products company in Colorado. Suddenly one day, he was given the assignment to take over operations responsibility for their struggling injection molding division. As ever, there was no increase in pay for taking leadership responsibility for an additional 125 people who exhibited many dysfunctions, like: high scrap, low morale, noticeable injury rate, low productivity, etc. (Aaron guesses the Doc would call these "the usual suspects" for the assignments he has had over the years).

The Doc remembers thinking about how all these career change, high-intensity, high-stress assignments were another interruption to his pleasant work life. He certainly would like to live without this frequent kind of interruption.

Delivering Safety Excellence: Engagement Culture at Every Level, First Edition. Michael M. Williamsen.
© 2021 John Wiley & Sons, Inc. Published 2021 by John Wiley & Sons, Inc.

However, rather than slipping into a BMW mode, the Doc recognized a lesson in life that all viable leadership needs to learn: "We get paid to be interrupted." What is more, these lessons continued to be an important part of what seemed to becoming his future work life, that of off the wall turnarounds. And the lessons were free (so to speak). Or as stated in another way, "The school of hard knocks is a very effective teacher, but the tuition can be quite high."

15.2 Management Reluctance to Be Involved

The leadership of the various functions at the main office of Doc's former company seldom spent any time at the frontline. They exhibited a classic definition of absentee management when they decided the solution to injuries was to pay for, and utilize, a famous high-cost behavior based safety (BBS) program. The Doc pushed back based on his experiences in the trenches when encountering this kind of program. He was very skeptical of supervision and hourly personal being required to evaluate (spy on) fellow workers against a checklist of compliance items. As he looked over the checklist, it had little or no correlation with the injuries that were occurring. He lost this battle, as the absentee (inaction) management did not want to get involved with any accountability for safety, or for that matter the poor operations they had spawned. Their inaction was what led the Doc to this new, challenging additional assignment. Upper management's solution seemed to be: throw some money at the problem along with an expendable leader who they can feel good about firing when he fails. (Aaron did not like the sound of this insight; it was way too close to home!) The employees dutifully took the BBS training and delivered the required observations, and sure enough, no real improvements in safety performance (injury rate) resulted. After a couple of months, the front office leadership stopped paying attention to the weekly observation reports, and the Doc let his people stop being involved in this non-value added BBS stopgap program.

As is the norm for turnarounds, weak upper management leadership nearly always seems to leave a number of areas where local leadership can begin making improvements. These "under the radar" areas take time to discover. However, small CI teams focused on local problems under the local leadership control have always become available after the Doc learned and defined the upper management consistently focused priorities (aka boundaries). Once again the Doc began kaizen teams involving hourly and salaried personnel who focused on all the production issues: scrap, logistics, cycle time, etc. However, each team also had the responsibility to resolve every safety issue, in every step of the particular problem process they were working on. With visible upper management involvement (the Doc), active middle manager involvement, focused supervisory performance, and active hourly participation the organization took about seven months to do a complete turnaround in all the usual suspects, including safety. World-class performance was within sight and the cross functional, all level participation teams were proud of their accomplishments.

15.3 Regulatory Audits

Out of the blue one Wednesday morning, an OSHA inspector showed up unannounced to evaluate the safety program at the injection molding business unit. The Doc interviewed her and contacted the absentee front office management about this day's abnormality. He then brought in a couple of hourly technicians who knew the processes and people, and he asked them to accompany the OSHA representative on her inspection rounds of the operation. She immediately threw a curve ball, at the end of her inspection when she demanded an audience with a dozen or more hourly employees, with no managers to be present. The Doc asked the technicians to pick out their compatriots who were interested in safety and have them attend this close out meeting. At the end of this secret meeting, as she left the facility, she commented to the Doc on how safe the operations were, and how the hourly employees took part in owning their own and their fellow worker's safety. She then unloaded on the Doc for her secret meeting becoming confrontational. It seemed the employees gave testimony to this being the safest place they had ever worked and why was she wasting the government's and their time when there were so many other target companies which were truly unsafe. The Doc smiled, wished her a nice day, and then went out to the floor to thank the employees and staff for taking responsibility to solve the many safety problems that had existed when he first came on board. You must always make time for regulatory reviews and complete their action items in due time. Work orders and occasionally CI teams can solve most of the issues. Once again, employee engagement in safety issue resolution also resolved the compliance and regulatory needs for the organization.

At this point the Doc stepped back and said it was time to introduce Aaron to another set of the simple "tools" people in an engaged safety culture use to develop excellence in all they do. Here was more food for thought and actions as Aaron and his committed leadership began their journey to world-class performance. As they moved forward these same tools would be personally experienced in greater depth as they were fit in with the other models and techniques used for delivering safety culture excellence.

15.4 Team Inclusiveness

Indeed, the place to start a safety culture initiative is not with the next safety program of the month. The place to begin is with the location(s) of the risks. That means including people from the most at risk work cells. Often there is a tendency to try and be all inclusive. Do not pollute the efforts with outside people who have little or no experience with facing the dangers that go with high risk work environment. Later, as an organization achieves success, and other administration or lesser risk occupations want to join in, have your SST lead them on the journey with teams made up of their volunteers who want to improve their own individual work area-related cultures. Keep each of the teams made up of frontline personnel that do similar work. In so doing, they can develop a culture focused on their specific issues and their people.

15.5 The Importance of Good Data and a Solid Improvement Process

Dealing with the information/data was next in Aaron's list of questions. Some centuries back the Greek philosopher Aristotle is purported to have said: "Treatment without diagnosis is malpractice." All too often, the Doc sees organizations deciding to follow the path of "Let's try this." These "dart throws" of the program of the month do not have enough diagnostic data, or strategic depth, to be successful. Over the years, the best diagnostic tool the Doc has found is Dr. Petersen's safety perception survey (Chapter 7). This quantitative survey provides multiple targets for the improvement teams to work on. The diagnostic information/data comes from "the voice of the customer," those who are doing the work and those who support the frontline work tasks and cultures. Often a worthwhile addition to the quantitative data is a set of qualitative interviews. In these discussions the "yes-no" questions can lead to good open-ended discussion inputs from the interviewees when they are given open-ended questions that go beyond the "yes-no" questions of the safety perception survey. Instead of the "yes-no" format the interviews focus on "tell me about" the various aspects of safe work, and the support thereof by staff personnel. These "voice of the customer" diagnostics provide an in-depth insight as to both what is good, and what is weak at the frontline, middle management, and the upper echelons of an organization.

Armed with real, credible diagnostic information, the RIW/CI team initiatives are provided with truthful information which helps them solve the safety culture weaknesses that occur daily in organizations. The Safety Steering Team (SST) (Chapter 17) then works through this data, as well an interpretation of the results. As a result the SST is tasked with putting all this material together to deliver the solutions necessary to resolve the issues of concern. The SST is comprised of hourly, frontline leadership and salaried personnel who are taught how to take the diagnostic data and information and quickly sort down to the needed issues and their respective priorities. Next the SST develops a multi-year plan to resolve the ***what*** and ***when*** of the issues they have decided to tackle as representatives of the company for the benefit of the company. With this information in place, the SST decides on the ***who***. Potential focus team personnel are discussed for the initiative, and SST members take the assignment to recruit these and others who would be good at solving the company wide problems being tackled.

The RIW/CI (Chapter 18) team receives training in what the Doc calls a Rapid Improvement Workshop (RIW) approach to solving culture issues. This RIW team is then:

- Given: survey and interview data.
- Trained: in simple but effective continuous improvement tools.
- Tasked: with developing their own charter, POP statement, and Action Item Matrix (AIM).
- Guided: through a multilevel process of how to solve the many problems, develop a training plan for the solution, communicate effectively with the SST and the workforce, and then run a pilot test of their proposed solutions.
- Shown: how to develop safety accountabilities for each level of the organization. These safety accountabilities are focused on what individuals must do on a daily or regular basis to reduce the possibility of incidents for the problem process they are working on.
- Tasked: with making a time line for developing and implementing the solution, including the various SST briefing reports, the pilot trial and the final roll out across the organization.

And thus, the 60–90 day in-depth process to resolve a target problem issue begins. Along the way the RIW team and the SST develop both the needed solutions and the needed skills to continue the RIW process, and focus on the relentless pursuit of a zero incident/at risk actions safety culture.

15.6 The Need for a Challenging Time Line

In the background a three-step data, development, and evaluation process has begun:

- **Year one**: The safety perception survey and in-depth interviews provide the necessary data to the teams. These RIW teams then guide and deliver the solution processes.
- **Year two**: A Safety Leadership Assessment (SLA) provides quantitative data on how well both the improvement process and safety culture improvement perception are progressing.
- **Year three** – the Culture Application Evaluation: Along the way there is a tendency to revert to what the organization people did in the past ("Why keep going

with all these efforts?"). After three years it is time to assess how well all the improvements and cultural transformations are being actually applied in the field, and with the various staff levels that need to be involved. Is the organization delivering a sustainable safety culture of excellence? Do the results and effects of the safety improvement process have a positive contribution to the overall costs/benefits/morale of the company?

There must be a noticeable pressure to keep moving forward in a limited space of time. The improvement process consumes expensive employee time and leadership resources. The processes and time frames explained to date (and in later chapters) have been proven to be very functional for safety culture turnarounds in the industrial workplace.

15.7 Urgency Followed by Complacency

The Doc reiterated some findings from his personal background in effort to highlight the death knell of complacency which can be a real danger that always seems to be lurking in the background. He mentioned that he began to become deeply involved in helping organizations achieve safety culture excellence back in the 1980s. At that point in his career, he was in charge of manufacturing engineering for a Fortune 20 company. One of the facilities experienced a fatality, which in turn led to a serious corporate-wide safety initiative. Within a week of the tragedy, the Doc was assigned responsibility for the safety of about 10 000 manufacturing employees who worked at 40 plants scattered across the United States. The organization ran an average recordable injury rate in the 20s and had no real safety focus. Production was king as measured by cost, quality, and customer service upstream indicators.

After some research, the Doc hired a famous safety consultant, Dr. Dan Petersen. For three years Dan and the safety focus team lived on the road developing a safety culture excellence model which quickly delivered a corporate-wide average injury recordable rate of about 1.2. The leadership and frontline personnel of this large organization were all very happy with a safety accountability culture that worked well across the entire nation.

After the initial success, everyone went back to their "normal day job." After all, they were convinced they had achieved "mission accomplished" and the Doc had left the company for other, greener pastures (so to speak). Some years later, after presenting the success story at a safety conference, the current safety manager who replaced the Doc for that organization stopped the Doc to discuss the organization's current situation. About five years after all had celebrated how good they were, the whole organization was back at about a 20 recordable injury rate.

The new safety culture stuck for a short time and then collapsed after the resources were reallocated to the ever dominant production culture.

The Doc's lesson from this: **you must never declare victory**. The pursuit of zero incidents/zero at risk activities MUST be relentless. If the organization backs off its safety emphasis, something like the second law of thermodynamics effect seems to take over (i.e. the whole system degrades over time). This is exactly what production culture counterparts have learned for their excellence initiatives in cost, quality, and customer service. All personnel must keep alive the push for excellence. To quote Yoda, "There is no try, there is only do." All people in the organization must remain engaged.

15.8 Series or Parallel Problem Attack Process

The Doc then threw Aaron a curve ball about other issues people on the journey of improvement through engagement of people often have to face and answer: How long must they stay engaged? What happens when we lower our injury rate? Does this process only work for high injury rate organizations? Out came a chart showing the journey of a heavy construction company the Doc and his group had worked with a while back (Figure 15.1).

High injury rate cultures typically have any number of unsafe conditions they need to address, the traps that are waiting to spring and catch (injure) employees. This company spent five years working through Level 1 problems on their own

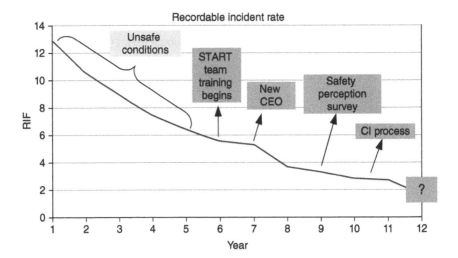

Figure 15.1 Past to future recordable incident rate.

until they hit a plateau where they had to use a new (culture) tool set. After checking the internet for resources they decided to begin with the Supervisor Training in Accountability and Recognition Techniques (START[1]) program. Next a new CEO focused on and became personally and visibly involved in safety. This was followed by their doing a Safety Perception Survey, which identified and detailed the safety culture weaknesses. And then the emphasis included an RIW/CI process on engaging their people in solving the problems they had been living with for a very long time. It took them 11 years to reduce their RIF by an order of magnitude (a 10-fold reduction in injury rate). If you take a look at the chart back in Chapter 12 you will note that with significant resources and focus it still took this Fortune 50 heavy equipment manufacturing organization an additional 13 years to reduce a RIF of 6+ down to 0.5. However, the chart also shows a five-year plateau at a RIF between 2.0 and 1.0. This stall in safety performance was finally broken once the company culture improvement tools were implemented. After the addition of an effective safety culture improvement paradigm, what would be considered by most to be an excellent RIF now continues to be improved upon every year. There is no sustainable quick fix for this organization (or yours) to save 65 000+ employees from serious injuries. It takes time, effort, engagement of all involved, passion, perseverance, and the utilization of excellent safety improvement processes and tools to accomplish this kind of a result. However, there is a lesson to be learned along with a much quicker way to reduce RIF. That is to begin the engagement RIW process immediately even for Level 1 (see Chapter 10) issues. Do not wait; tackle the reactive trap issues by engaging your hourly and salaried leadership on day one and thus build and live the attributes of an excellent safety culture. This process must not become one of working on each of the levels in series (one after another, waiting to begin work on the next level after the current level work is completed). This safety culture improvement approach is meant to be run in parallel (Level 3 CI/RIW teams working on the safety culture weakness problems the Level 2 diagnostics pointed out at all levels).

As an example of this parallel approach the Doc showed Aaron a RIF chart of roadway construction company that began the entire safety culture engagement process on day one (Figure 15.2).

This approach took the organization only four years to go from being an industry laggard to becoming an industry leader.

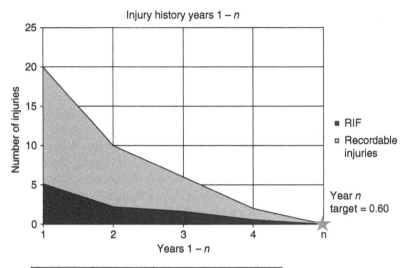

Year	RIF	Recordable injuries
1	5.12	20
2	2.2	10
3	1.64	6
4	0.58	2

Figure 15.2 Lagging indicators.

15.9 The Importance of Viable Metrics

With this timing concern addressed, the Doc added one last "food for thought" item. This new input needed to become a part of the detailed work Aaron and his team would learn as they engaged their organization in their journey to safety culture excellence. Working to resolve the real problems and evaluating the culture improvement process has the teams delivering Key Performance Indicators (KPIs). KPIs are practical leading indicator metrics that measure day-to-day progress against a series of sustainable goals. The KPIs help leadership evaluate how well the important parts of the culture are doing. The KPIs are necessary to an organization in their learning and delivering sustainability of excellence in safety. They are necessary outcomes of a deep dive process in what is needed to deliver world-class performance. Some organizations use different terms for KPIs, like metrics, or a dashboard. No matter what the term or language used, these are the critical few items that must be consistently delivered if the organization is to achieve excellence.

> *Aaron nods his head as he is understanding the what, when, who and how of the new culture life he must lead. The Doc then tells him about his meeting with the board of directors that day. It was an intense group which included the organization's; legal counsel, new CEO and, to the Doc's surprise, a board member who was an under classman at the same university where the Doc attended. The Doc presented the data and the proposed improvement plan strategy in a no holds barred, no time limit, intense interactive dialogue. The bottom line was that the group seemed in agreement. They now await a proposal from the Doc's team. Depending on all the uncertainties, the Doc expects to see Aaron in a month or two. So get ready Aaron, your time of working your way out of the barrel is about to begin!!*

Note

1 Original publisher CoreMedia Training Solutions.

Part III

16

Moving Forward to Safety Culture Excellence

*Aaron, like all who want to move forward with a process critical to their orga-
nization, struggles with the inevitable waiting period required for an initiative
to slog through even a relatively small bureaucracy. He talks to all the critical
decision makers more than once and continues to wait. Aaron mentally goes
through the material they have dialogued about during the Doc's visits and
is somewhat comfortable with all this background information. However, the
personal frustration of the delay in getting good and logical ideas transferred
into concrete action plays mind tricks on Aaron. "Might as well go back to the
old reliable solution," so Aaron picks up the phone and calls the Doc. He starts
off with the logical question of what comes next when the signatures are inked
and the process of improving a sick safety culture kicks into gear?*

The Doc begins the explanation with a story about when he was contacted by an
offshore oil company with large petroleum reserves. They had just experienced a
multi-fatality event that truly shook the company to their core. A contractor was
doing a hot tap on a large diameter crude oil line when the crew made a mistake. The resultant explosion resulted in a large number of fatalities. In fact, the
inferno was so intense that the Doc was told they could not even find the ashes
of the victims. In addition to the fatalities, the fire also caused the shutdown of
their tanker loading facility. The company arrived at a realization that their safety
culture had to go way beyond merely focusing on the fundamentals of regulations
and compliance.

After the immediacies of investigations and the resulting inevitable fallout, the
time arrives for the next steps to implement a safety culture turnaround. The Doc
explained this process is an in-depth safety culture improvement approach which
evaluates an organization's safety culture strengths and weaknesses and then leads
to a process of significant improvement.

The first step is for the organization to take the safety perception survey
(Chapter 10) from the top management level all the way through the organization

Delivering Safety Excellence: Engagement Culture at Every Level, First Edition. Michael M. Williamsen.
© 2021 John Wiley & Sons, Inc. Published 2021 by John Wiley & Sons, Inc.

to the frontline workers. The safety perception survey gives a snapshot baseline of safety culture reality and is the voice of the customer (all those employees from the bottom to the top of the organization who took the survey) when the improvement work begins. Once the survey has been run and the data interpreted, it is time to share the results with upper management. The diagnostic data and interview results lead to interpretations and recommendations that are presented to leadership. Obtaining the decision by upper management to proceed is mission critical, as are the follow-on actions that the decision entails.

Organizations decide who receives this report of survey results. Is it only upper management, does it include the board of directors, and what about frontline employee participation? The reality of frontline to upper management barriers is a culture checkpoint that leads to the decision of whom to include in the day-long leadership round table. This event provides an in-depth review of the safety culture, as well as how to improve and deliver a culture of safety excellence.

Hopefully, it is principled leadership and not a tragedy which brings this group together. However, it seems that many organizations continue on with what they have always done in the safety-related processes until "the pain of remaining the same finally exceeds the pain of significantly changing what they are doing." The book *Good to Great* by Jim Collins[1] explains the plateau of good enough being the enemy of great for this kind of organization. Unfortunately, there always seems to be some kind of ongoing crisis situation or the ongoing whirlwind of activities which pull away resources and focus from doing something about eliminating injuries. There is an uncertainty of how a zero incident/at risk activities safety culture can be developed in an organization which has very little understanding of safety beyond the basics of the compliance and regulatory reaction approach. These are formidable barriers to overcome. That is, until the next tragedy occurs.

The Leadership Report Out (LRO) addresses all: the uncertainties, the fears, the data, and the approaches to achieving solutions. In a six to eight hour session with the leadership team:

- The survey background, the organization's data and the data interpretation are explained in detail.
- An in-depth explanation of safety culture realities is presented, which includes an interactive discussion of what really causes injuries and incidents (Figure 16.1).

Why do incidents/accidents occur? Much of the safety literature points toward 90 + % of injuries being the result of frontline employee mistakes. These results can often lead to a focus on punitive measures for the unfortunate injured employee(s). However, there is a different answer which looks us straight in the eye when we consider the diagram below developed over the years by the Doc and his customers: what causes at risk behaviors by the frontline people? The answer is not

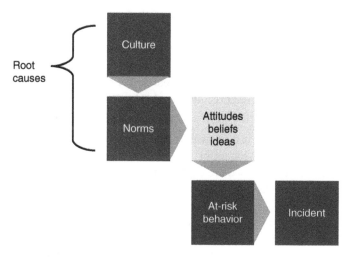

Figure 16.1 Why injuries happen.

stupidity, or laziness, or cluelessness. Our employees do not come to work with the attitude that today is their day to get injured!

The apparent root cause is the ideas, beliefs, and attitudes that "this activity is what we do on this job for this organization." It is who we are. Digging a little further into the root cause ladder, we come to an understanding that this is the norm for who we are and what we do. And it is not just the frontline employees with the attitudes, beliefs, and ideas who think these actions are ok. The message comes from further up the ranks as well: our supervisors, their middle management leaders, maybe even top management. The Doc related a story about being asked to bid on safety services for a large heavy construction organization which had won a billion dollar plus contract for a significant building complex. As the Doc's team interviewed the construction company's CEO, the CEO informed them that it was dangerous work and yes, no doubt there could (would) be one or more fatalities associated with this complex, dangerous contract. With that kind of expectation the Doc and his team declined to put their name on the bid list. They had no desire to becoming entangled in the CEO's and his company's poor safety culture ideas, beliefs, and attitudes.

Another step up in the root cause analysis diagram indicates that these weak, improper, dangerous ideas and beliefs are really the culture of the organization. It is who they are, what they do, and how they do what they do. Punishing, blaming, and complaining about frontline worker ideas, beliefs, and attitudes does not solve such root cause safety culture issues. In all likelihood, this culture cancer would be growing ever stronger as these same ideas, beliefs, and attitudes are continually reinforced for a number of years.

The safety perception survey and field interviews help the leadership start out with a diagnosis of the organization's true safety culture. As Socrates is rumored to have stated a few thousand years ago, "Treatment without diagnosis is malpractice."

This is often a part of the explanation given for weak safety performance during a Leadership Report Out (LRO) at the executive level. Also included in this LRO are such things as:

- The safety accountability model is discussed in detail along with Dr. Petersen's six criteria of safety excellence (Chapter 9).
- The six levels of safety performance are explained in detail so the leadership understands the logical progression of improving the company's safety culture (Chapter 10).
- A brief, simpler diagnostic is run with this group of leaders. When they fill out and discuss this short interactive diagnostic, it allows them to determine their organization's safety performance from their perspective based on the data, interpretation, models, and discussions which have taken place.
- The continuous improvement (CI), Rapid Improvement Workshop (RIW), and Safety Steering Team (SST) models are also explained in detail. As a result, the leadership team gains the knowledge on how to engage themselves and their people in developing a zero incident/at risk activities safety culture.

When the LRO is completed, it is time for the organization's leadership to decide whether or not to engage in the Continuous Excellence Performance (CEP) process and initiative (Figure 16.2).

CEP Explained

The Continuous Excellence Performance (CEP) process is a continuous improvement methodology which enables an organization to transform its

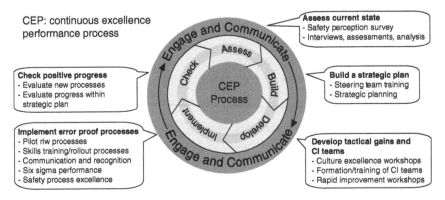

Figure 16.2 A process to achieve safety culture excellence.

culture toward excellence. It was developed by the Doc, his associates and Brad Cosgrove as they worked with a number of their customers on this concept. In this context, the purpose of it is to focus primarily on safety excellence. However, it can be used to improve any process within an organization.

The CEP process is comprised of five phases with communication and engagement activities interwoven throughout. It is grounded in three core principles:

Commitment

For the CEP process to be effective leadership commitment is critical, from the top leader through the frontline employees. This approach aligns with Dr. Dan Petersen's first three criteria of safety excellence:

- Top management must demonstrate visible commitment.
- Middle management must be actively involved.
- Frontline supervision must be performance focused.

Leadership must consistently demonstrate commitment to safety by integrating safety activities into the regular business conversations at every opportunity.

Accountability

No transformation effort can be successful, nor can improvements be sustained over time, unless new activities and expectations are built into the organization's accountability system. CEP accomplishes this by including Dr. Petersen's expanded five steps to accountability in the Rapid Improvement Workshop (RIW) process. When a team strives to improve a process, they build in the five steps of accountability at every level of the organization:

- Define the work to be done.
- Train/educate/confirm the employees know how to do the job safely and correctly (see Chapter 17).
- Resource the efforts needed to do the job safely and correctly.
- Measure how well the employees do the tasks.
- Provide proficiency feedback to the employee.

Much of the leaders' activities associated with such a new process describe what the leader must do to hold their subordinates accountable for performing the subordinates' defined activities. This holds true at every level of the organization.

Engagement

People are much more likely to support what they help to create. CEP requires engagement at all levels, but it focuses especially on frontline employees. This is normally where the most dangerous hazards reside. It is the employees at the frontline who must be influenced the most. The frontline culture is where the most hazards and at risk actions typically exist. Through the continuous improvement teams, employees participate in developing solutions, appropriate accountabilities, and improvement opportunities for the problems identified in the assessment phase. Through effective engagement, employee attitudes, beliefs, and ideas are influenced so that working safely becomes the new culture. Dr. Petersen's last three criteria of safety excellence are all put into practice through the principle of engagement:

- The employees must be actively engaged.
- The safety system must be flexible.
- The safety system must be positively perceived.

Beyond the six criteria of safety excellence, as noted in the graphic above, the five phases of CEP are:

1. Assess
2. Build
3. Develop
4. Implement
5. Check.

Within each phase are elements which must be executed with excellence. In order for this cultural transformation to occur, effective communication and engagement are required throughout the process. In the "check" area of the above model are adjustments made in the "Build" and/or "Develop" areas as a result of observations in the "Implement" phase? A general factor in the process as explained above is keep going around the circle, i.e. the continuous improvement cycle.

With an engagement decision, a specialist *may* need to be brought in to help the organization set up and train the SST and the RIW teams. A schedule is also developed to firm up the engagement of the hourly staff and frontline supervision, upper management, safety and communications personnel who *ALL* will be involved in the processes.

> *Aaron has a better understanding of what comes next. Added to this is his observation of how the Doc has already interacted with the executive staff, the board of directors and the frontline employees. Aaron wishes he had the Doc's experience and ability to effectively communicate. The Doc wishes he had Aaron's youth and exuberance. After the report out, the real work by the teams begins, and this phase concerns Aaron because their performance to date has been nothing short of miserable. How is the Doc going to herd all the cats which have resisted any kind of leadership in the past?*

Note

1 HarperCollins Publishers Inc., 2001 ISBN: 978-0-06-662099-2

17

The Critical Safety Steering Team

> *Aaron would much rather herd horses than cats. Many of the 'cats' in his orga-*
> *nization; scratch, bite, hiss, and have bad tempers. There are more than 1000*
> *of these critters in Aaron's company. The Doc better be really good, because no*
> *one in Aaron's company has been able to do this before, as the injury statistics*
> *testify. The Doc's on site work coming in two weeks is going to be an event to*
> *behold. Aaron sure hopes he will observe a process that brings success, not just*
> *another scattering of felines and cat fur.*

Before he left at his last on-site visit, the Doc brought together the executive group and talked about a Canadian pipeline company that merged with a smaller competitor, which added a critical ancillary technology. The resultant company was much stronger than either of the two entities that combined their forces. However, there were significant differences in their two safety programs and approaches. The new CEO was very much committed to the safety of all the employees, but unable to get the combined company's two separate safety leaders to come together and connect the dots. After the Doc met with them, they agreed to have all their employees take the complete safety perception survey. During the subsequent Leadership Report Out (LRO) and cultural data discussion, the CEO, the two safety leaders, and the senior staff met together. Their true differences, including both strengths and weaknesses, were clearly apparent and the Continuous Excellence Performance (CEP) (Chapter 16) process was deemed to be a good fit for their culture of always staying focused on improvement. During this Leadership Report Out (LRO) they discussed and decided to start with a leadership team which would guide their efforts across all the divisions and individual sites to the common goal of achieving a zero incident/at risk actions safety culture. They also agreed on the membership of the guiding coalition Safety Steering Team (SST).

The Safety Steering Team is a critical success factor in moving from data, into tactics, and then to a strategy for sustainable execution. The combination of mix,

Delivering Safety Excellence: Engagement Culture at Every Level, First Edition. Michael M. Williamsen.
© 2021 John Wiley & Sons, Inc. Published 2021 by John Wiley & Sons, Inc.

meetings, time frame, strategy and tactics, developmental training, POP, charter, etc. then becomes a viable business approach to solving an organization's problem of weak performance/culture in safety. This unit operations process is also a key performance indicator for developing strong safety culture leadership from the top through the frontline of an organization.

Steering Team Overview: A strong and capable steering team is the key to effectively executing the CEP process and sustaining improvements made by CI Teams. Once established, the steering team will remain the strategic guiding force for safety management and transformation into perpetuity. The steering team training and strategic planning session is only the beginning of the work for the steering team. Most CEP initiatives include a year or more of steering team meetings for the team to attend, learn, and perform. The ideal number is 12 in the first year…one per month. Each steering team meeting requires most of a day. Any time remaining short of a full eight-hour day is used to connect with senior executives and other company decision-makers to discuss project updates, things going well/employees to recognize, and issues of concern. After the first year of a CEP work and interaction, the Safety Steering Team should be well educated on safety excellence and what is required to create and sustain the safety program. At this juncture the SST members will also have a working knowledge of all the models and tools required to enable them to succeed in their organization's efforts. The SST leadership and members make a commitment and take responsibility to ensure this outcome in year one.

Membership: The steering team should be composed of people from a cross section of the organization from all levels and areas…typically a total of 8–12 people. In large organizations with many divisions or areas, it may fill out to 12. More than 12 members for this strategic leadership team typically becomes problematic for decision-making and problem-solving. The objective is to ensure that people in the organization feel like they have a representative at the strategic planning/decision-making table. The Safety Steering Team delivers the representation at that table. This steering team should be comprised of the following employees:

Executives – one or two executive leaders who report directly to the top leader of the organization.

Middle managers – one or two members of middle management.

Frontline leaders – two or three frontline leaders.

Frontline employees – two or three hourly employees from the frontline.

Safety director or manager – the key safety leader in the organization; typically present as an expert resource, not necessarily a voting member.

Administrative support – a member of administration who will be responsible for recording minutes and keeping the AIM (Action Item Matrix) up to date, not necessarily a voting member.

Other roles – the 8–12 total membership is a guideline for the abovementioned SST. Some of those participants may have to take on other roles in the team for the team to function effectively. Those roles include:

- Team leader – selected by the steering team, works to plan the agenda, communicates with team members as necessary, chairs team meetings, oversees AIM activities, and serves as liaison with the executive team along with the executive sponsor.
- Executive sponsor – removes obstacles, ensures needed resources are provided, demonstrates visible upper management commitment, and represents the team's interests in executive team meetings.
- Communications chair – ensures all necessary SST communications are developed and delivered, helps CI teams with communication needs, acts as keeper of the communications strategy, and works as the liaison with the organization's communications people.
- Training coordinator – helps to execute monthly training for the steering team, acts as liaison with the company training department, supports the training plans that come from CI teams, helps ensure all training logistics are executed effectively, and ensures new team members that are added are properly oriented to the team.

Terms of service – typically, members serve a two-year term, rotating one half of the team each year to maintain experience and continuity. However, the Safety Steering Team must decide how they want to establish the rotation of members. During the startup phase, half the membership will often serve a three-year term in order to establish the rotating cycle. Each time new members join the team, experienced members must ensure the new members are oriented and brought up to speed. This process requires that a new member orientation plan be in place as a part of the SST charter (see Appendix C).

The Safety Steering Team frontline members also provide critical feedback in their work teams as to upper management engagement and commitment required for the frontline to get to zero. This arrangement establishes a critical credibility and trust relationship to the frontline: are they trustworthy and believable or phony? Once the SST comes together, there is no place to hide phony. It is very important to be careful as to who is chosen for this one-year plus ongoing assignment in achieving organization and safety excellence.

An often asked question is, whether the SST chooses the initiatives to work on by Pareto vote (see below), or does the SST go back to the organization's people and ask them to decide the focus for their efforts. The best answer is to rely on the results of the survey and interviews already completed. These show the organization's decisions as to what needs to be worked on. It is now time for this SST leadership group to take action. Do not wait another "X" months to go through the feedback process all over again, take action now! Remember that time is money including what has already been spent on past survey and interview work.

The SST is tasked with implementing a "complaint equals goal" strategy, not more BMW about what is wrong. No more "shoulding" on anyone or looking to others to solve your problems. The SST membership counts on the government and other outsiders for absolutely nothing. If it is to be, it is up to this team to deliver a culture that solves its own problems: "You must fork your own broncs."

In explanation: both the past and the current culture of living on a farm or a ranch is that of self-sufficiency. If your horse is hungry, no one else is there to feed it. You have to fork (feed hay to) your own broncs (broncos or horses). For many years the Doc earned a living helping organizations escape bankruptcy and become profitable. Obviously improving safety was an important part of the turnaround necessities. These sick companies frequently (continually) looked for someone else to solve the problems they had created. They were trying to live in a world of "The government should do... The union should... The management should.... The engineers should..."

This never worked. The "Theys" of the world had to become the "us" who had the problem and needed to develop viable solutions. An organization has to learn to fork its own broncs. To paraphrase Michael Topf, a contemporary safety leader of the Doc, "You shall not 'should on' anyone." The expression the Doc used in these circumstances is, "Count on 'they' for absolutely nothing." In other words, what can **we** do within **our** own resource limits to solve the problems **we** have? Instead of looking for excuses that keep us from solving problems, we need to list the barriers we face and begin breaking down those barriers with the resources we can control. From a safety profession perspective, the Doc has painfully learned that the leadership must count on the many and various OSHA organizations for "absolutely nothing." The corollary to this is to be thankful if any of these outside resources do actually show up and assist.

The SST and the RIW teams often struggle with what problems to tackle first. They cannot fall back on the old engineering approach of creating solutions – solutions designed to solve the wrong problems! Should we start with the hardest issues, or the easiest, or where the passion exists? Having the SST members use Pareto voting will solve the what-to-do-first question.

Vilfredo Pareto was an Italian economist and sociologist who is known for his theory on sorting data down from many events to the relevant few that make a difference, i.e. roughly 80% of the effects come from 20% of the causes. The Doc explained that he simplified the math to a "Pareto voting" process. This allows a group of frontline and management employees to quickly take all the safety perception, interview, and field complaint information and decide on the few items they will focus on with the RIW teams. In "Pareto voting":

- All the problem issues are written down/listed in front of the group on a white board or multiple flip charts.
- Each team member then gets three votes.
- One at a time, each team member goes to the list and votes by putting a mark next to the item he/she thinks the team should focus on. Only one vote per item and the team member tells why this is important enough to be a focal point. Other team members may make comments as to the choice.
- This person by person "Pareto voting" continues until all members have voted and discussed their choices.
- The four to five high vote items are then discussed by the team and a final priority list for the RIW teams is decided upon.
- All members have participated in this sort down process, no matter their position in the organization. The open-ended discussion for each item is a powerful approach to delivering a workable decision.

However, when the SST team members review all the issues before them there can still be a sense of being overwhelmed. Can we really shovel through all this stuff and deliver a viable, sustainable zero at risk action safety culture? Two guiding principles are usually sufficient and effective in conquering this fear of the unknown:

- The first comes from the movie *Apollo 13* in which Jim Lovell cautions his crew to focus on tasks at hand rather than worrying about what will happen way down the line. He says, "A thousand things have to happen, in order for us to get home safely. You're talking about number 876. Right now we are on number 8." Too often we think we need to know all the details before we take the smallest step. In this safety culture improvement continuum, the team's job is to do the best next thing. (Another good axiom to remember is that prescribed by Yogi Berra: "When you come to a fork in the road, take it.")
- The second guiding principle is out of the Doc's past maintenance engineering days where there was a seemingly unending string of breakdowns and the resultant stress on all team members. The Doc's advice to his crews was, "Illegitimus non carborundum" (do not let the harassing circumstances and people wear you down). Stay in the game, it takes 9–12 months to begin seeing culture change.

Long-term commitment is required. It took decades to dig the poor performance hole the organization is in. It will take years to build a culture that just naturally fills in the poor performance hole on a day-to-day basis. Wrestle with the demon that is facing you. Do not be worried about all the possible "What if?" demons, most of which will never come to pass.

As time progresses, another characteristic of an SST is the strategic leadership team expands its efforts to other areas which need to be improved, e.g. customer service, logistics, maintenance, etc. In this manner they continue to engage the people with their minds and efforts in the organization's relentless pursuit of excellence in all that they do. But only **after** a culture of safety excellence is established. If leadership branches out too soon, the safety effort withers on the vine as the tyranny of the urgent pushes all else aside.

There is an additional bone of contention scenario which frequently rears its ugly head, a debate about:

- Is zero really possible?
- What is the zero incident rate reality?
- What is a good incident rate goal?

Whenever this zero injury rate topic surfaces there seems to be an emotional sparring that goes on. Is it futile to set zero as a goal versus accepting the inevitability that somewhere along the line someone will get hurt. Is a zero incident demand like living in a fantasy world?

A few careers back, when the Doc had responsibility for manufacturing engineering, an almost identical debate seemed endlessly waged over whether or not an organization could achieve zero downtime. Likewise, it was both emotional and agenda driven by the sparring combatants, who never really gave into, nor accepted their adversary's premise, no matter what was said. This caused the Doc to come away with a threefold view point – dark, light, and gray:

- Dark: Bad things happen and you just need to do your job diligently and accept what comes your way.
- Light: No matter what happens people will strive for delivering excellence with every tool they can utilize.
- Gray: Why debate this inane emotionally packed issue? Just go do the best you can and live with what comes your way.

In that maintenance world and the striving for zero downtime career the mantra became:

- Can we achieve zero downtime for a day?
- For a week?
- For a month?
- For a quarter?
- **For.......?**

This was the continuous improvement challenge mindset, **the relentless pursuit of zero** downtime. This same mindset must be the rule in safety, i.e. the quest for no injuries to our living assets (our fellow employees both on and off the job):

- What did we learn from developing permanent fixes?
- How could we apply these lessons to similar issues?
- How could we apply these lessons to those issues that were more loosely aligned?

The end result of this zero focused mindset was to develop and reinforce a culture that kept pushing the envelope, in a relentless pursuit of a culture of zero downtime and zero at risk activities which everyone in the organization lived and supported.

This then is a similarly good read on the endless zero injury/zero incident rate debate. What good does it do to set a goal and then celebrate the achievement of a "satisfactory level of injuries?" You cannot possibly win this kind of debate. Why not instead begin leading a culture which keeps achieving the next threshold and then moving that achievement to the next level, relentlessly?

Using the SST model, a different kind of corporate safety guidance is delivering a strategy and results beyond what the traditional compliance-based corporate group has typically been able to achieve. A corporate level Safety Steering Team, with a focus on safety culture strengths and weaknesses, brings together frontline practical personnel, safety professionals, and upper management. The Safety Steering Team process engages corporate and frontline leadership, safety culture survey and interview data, continuous improvement teams, and emerging safety needs. The combination of corporate support, frontline practicalities, safety culture reality, and effective solution techniques delivers rapid, workable solutions, which include functional safety accountabilities. The resultant zero incident/at risk activities safety culture intensity spreads and takes hold at every site across an organization.

Strategic Initiatives – This corporate level Safety Steering Team approach focuses on strategic initiatives, rather than reacting to the latest injuries (being proactive rather than reactive). The team itself is also very different from the reactive safety culture committee norm. There are available corporate leaders at the VP level and a corporate safety resource. However, the rest of the team is comprised of frontline responsibility personnel who volunteer to assist in proactively eliminating the cause of injuries, no matter where, or why they might occur. The team also has observers and resource personnel like communications support. These additional people are necessary to provide consistent, timely safety improvement messaging to the various workgroups throughout the organization. There is usually an available midlevel line/operations manager as well as a couple of frontline supervisors and a couple frontline hourly employees. In other words, more than 50% of the team's membership comes directly from frontline responsibilities. This structure ensures the day-to-day practicalities are a part of every improvement initiative. Frontline employee engagement is also critical in breaking down barriers and potential trust deficits which commonly exist between management and production labor.

The SST meets monthly, usually at a corporate headquarters facility. Here they briefly review the state of the company and the state of its safety culture. In doing so, frontline personnel on the team are updated with respect to what is going on at a corporate level. In this way, they are looked upon, and sought out by their peers back in the field, as credible, up-to-date resources for information about the company.

The Safety Steering Team then moves on to what is being done by continuous improvement teams they have sponsored and launched. Their chartered teams always include a steering team member or two, which makes the progress briefing current and crisp. The monthly agenda also includes safety training which provides in-depth knowledge on foundational safety material, such as incident investigation, near miss/close call reporting, effective frontline safety communication, how to lead safety continuous improvement teams, etc. In this way the steering team members become viable, knowledgeable safety champions during their one-year plus tenure on the steering team. In turn, as they rotate back out into the field, they provide greatly increased knowledgeable safety leadership beyond that of the usual safety resource personnel. This system results in a noticeable increase in engaged, credible frontline safety commitment.

It is important for the steering team to meet on a monthly basis. Doing so creates consistent pressure to analyze, focus, and execute ongoing safety improvement activities, which keeps the safety culture improvement initiative efforts from dying on the vine. The meetings are moved to field locations once or twice a year so that upper management support of safety is visible in the field. This kind of Safety Steering Team strategy provides guidance, resources, noticeable upper management support, and active frontline participation. The result is a highly visible, active, and relentless pursuit of a zero incident safety culture.

In troubled organizations, this process of taking responsibility for daily and regular preventive actions sets the stage for solving the problems which must be conquered. To achieve this kind of culture requires a team approach which gets the organization comfortable with using all its personnel resources. The "Lone Ranger" approach does not work. An organization needs a team of dedicated people, with diverse talents, relentlessly tackling the problems which it faces. And they must take time to recruit and put together the good functioning and talented team that can conquer the many problems of the past and present. It takes time to build the team, to learn the skills, to solve the next wave of problems, to operate "autogenously" (an engineering term for self-sustaining fire). Just like on the farm, it takes time to learn how to fork your own broncs, and all the other activities that are necessary to survive and excel.

Aaron nods his head in understanding, but then wonders what the safety broncs are that he and his company themselves need to start forking and not waiting for others to do so? The Doc smiles, and immediately emails him the list that resulted from the interview process:

Interview Report – As mentioned above, the data/information gathering phase includes a safety perception survey of all personnel. There is also one-on-one "tell me about the following issues" interview responses from a smaller sample of selected people. This information is analyzed and condensed into an interview report which summarizes the salient points of what the organization's people believe is the true safety culture that exists. What is shown below is an excerpt that represents typical material from such a process.

- There is a frustration with a growing dysfunctionality which has both new and seasoned employees seeking other employment in the hot labor market in the surrounding areas. The employees want someone in leadership to point them in the correct direction and help them get there. Across the organization there is a true hope that its leadership will be able to pull this organization back together after it has struggled with safety culture issues for more than a decade. There is also a genuine fear that the blame game may very well constrict leadership's ability to dig the organization out of the deepening hole that has been worsening for years. There is a frustration about the organization not using its brain power of the excellent degreed and non-degreed people who work for the company.
- Strong cultures of both safety and belonging do exist within a number of departments, and at some remote locations. Bright spots of good engagement and leadership across the hourly and management ranks can provide opportunities for learning and thus can be replicated more easily elsewhere with diligent effort. Many of the people throughout the organization want to "develop a culture that they are glad to work for and retire from" which seems to have melted away over more than a decade of time. The desire to improve safety genuinely exists. Interview feedback suggests that over time there has been degradation in many of the organization's cultures. A number of hourly and salaried people want to help build a strong, viable organization which will be needed for the next 50 years. However, there is also an ineffective leadership vacuum to do so in the hourly and salaried ranks. This is an obstacle that must be overcome.
- The training process for apprentices and others is a loosely defined legacy/storyline approach, administered by current journeymen or other existing employees. There appears to be little or no documentation as to what is taught and the teaching methods used. Nor does there appear to be sign off/certification of proficiency process for the apprentice, or new employee, or even the assessor. Safety culture improvement training has occurred to some extent. However, with no follow-up, accountabilities, or outcome expectations, this training investment in improving safety culture appears to have had little or no positive effect. Keep in mind there is a difference between mere training and educating. The classic example is that of Pavlov. He could train his dogs to

salivate (a conditioned response), but he could not educate them to think and pursue a purpose. The goal of education is to develop thinking skills!

- Upper management safety accountabilities are seemingly nonexistent. The current definition of accountability is more that of discipline/punishment. Instead, the definition of accountability needs to be a culture of doing the job safely and correctly with ownership of the results by those who are doing the job. Any field presence by these managers is extremely rare. Whenever frontline leadership needs upper management support, the management shifts to their need/desire/requirement for significantly too much justification being required for any positive actions. In turn, this malaise leads to no real action and a secondary degradation of an already poor morale throughout the workforce, with a lack of trust at the front line. If there is a cost involved to satisfy the frontline needs, the crews believe the issue will neither be addressed nor completed.
- The frontline crews had difficulty naming or describing good management role models. Their opinion was that a lack of experience, a fear of making the wrong decision, and some very weak safety attitudes/abilities from the office staff personnel exist every day. This has reinforced a frontline check in the box mentality of filling out another "worthless safety form" which delivers no action. The end result of upper management lack of accountability is enabling/supporting a culture at the frontline which also lacks accountability. There is a spirit at the frontline that: if excellent safety was honestly supported by upper and middle management the frontline would gladly engage in delivering this same excellence. It will require some hard work and long-term engagement by all parties to overcome all the years of mediocrity being the default safety culture. There needs to be a genuine code of excellence for management and labor that is supported, followed, and lived by everyone, every day.
- Fear of discipline is near universal for field hands. As is a norm for this kind of weak safety culture approach. Actual discipline takes a long time to develop and at a much lower frequency than what the field hands suspect. Nonetheless, the fear leads to a frontline culture of:
  ○ Reduced incident reporting.
  ○ Little to no near miss/close call reporting.
  ○ Closed safety meetings by crews who will not carry on a safety discussion when a management person is present for fear of a "mole" in the room.
  ○ A rumor mill mentality that does not get to fact.
  ○ Long delays in addressing real safety analysis, issues, and occurrences.
  ○ Hiding of the truth.
  When all these factors and other safety sidebar weaknesses are added up the end result is extremely poor workforce morale and ongoing high injury rates.
- Across the organization safety meetings are generally viewed as a boring waste of time which must be endured on a regular basis. Jobs are not expected to be

done well, or particularly safely, just good enough to move on to the next series of tasks. Management leadership seems to have little or no interest or knowledge of annual total costs of claims. The injury claims routinely amount to more than $1 300 000 annually, with an associated personal pain and suffering of 90 recordable injuries and 39 lost time injuries. The effect of poor safety performance and a lack of culture improvement and its associated degradation on any of the organization's management leadership annual reviews is minimal to nonexistent.

- There is a belief that there are too many committees which do not make decisions or fix things. Even with all the joint labor management meetings, there exists a feeling that notable important decisions are often made with little or no consideration for hourly input or teamwork. The safety work order system/process is broken: things just do not get done. There is no consistent, effective process or leadership follow-up from these meetings, which only serves to further degrade employee and organization morale. This entire process has led to feelings of "why bother?... nothing gets done... it's another program of the month... things just peter on ......" Or the other common approach of listing pages and pages of what needs to be fixed with no associated approach of how to fix it. There needs to be both a "what" and a "how." As items do not get done, or are not reported as being done, the festering complaint culture and ever worsening safety culture only get worse.
- An "Us versus Them" mentality exists between management and the frontline employees. Neither management, nor labor view management as a part of the safety team. Management frontline presence is all but nonexistent. In fact, there is at least one line responsible management person who agrees management has no idea what happens in the field. This helps deliver a feeling that management lacks:
  - Visibility.
  - Credibility.
  - Engagement.
  - Trust.
  - Any real commitment with respect to safety.

  Crews do their own thing, as best they can, without positive reinforcement from above, and with a real fear of punishment if an injury or incident occurs on their watch. The end result is a culture of merely being good enough, rather than excellent, because mediocrity is treated the same as excellence. The organization is not taking advantage of the knowledge and experience the hourly and salaried people possess.
- Safety incident analysis bogs down as each incident is believed to be handed over to Employee Relations (ER) for a review of potential discipline. In turn, this leads to little or no root cause analysis or lessons learned which can be shared and applied with field personnel and leadership, in any efforts to eliminate future

incidents/injuries. A common hourly evaluation is that management will take three to six months to slog through this broken process before there is even a slim possibility of any necessary forthcoming fixes. There is a thread belief that real training is necessary at all levels on how to work effectively with one another.

- Leadership development is a significant need in the present and the future. As seasoned personnel age, and inevitably leave, there is a necessity to develop consistent, sustainable safety leadership strength at the frontline and in upper management. The common urgency mindset focus on the immediate in the organization seems to have erased thenecessity for a mindset about the need for developing strong leadership that can sustain a more excellent organization in the future. At present, a tug of war is going on about how to achieve respect. Is it leadership verses management, or command verses demand, or anything that can be done that leads to respect? The end result of this struggle has everyone, hourly and salaried, on the defensive. This is one of the organization's notable dysfunctions and frustrations which has a number of people feeling desperate about the future.

- Instead of using safety improvement to bring labor and management together, poor to nonexistent leadership of this process acts as a deterrent to morale, safety performance, reduced cost, and any other necessary improvements to the organization. With no real improvement intensity on safety, the default position of operation's performance focus continually takes precedence over safety needs, no matter how much lip service is given to safety being number one. The result is that the approach to safety continues to be reactionary, rather than proactive.

> *Aaron chokes, shudders and thinks maybe he will actually choke in the process of forking all this stuff to his ever reluctant bronc of an organization. "Holy Mackerel, what a list of dysfunctional realities!" The Doc senses his shock and promises to be his sidekick during the process, which starts in two weeks at Aaron's ranch (company). The Doc leaves him with: "Hang in there trail boss, we are going to be successful. The broncs and the cats are all lining up for the round up!"*

18

The RIW Process

Aaron joins the safety steering team (SST) for some initial training on the Rapid Improvement Workshop (RIW) approach to solving the many problems that are running rampant in the organization. After introductions and small talk, the Doc dove into training the team of executives along with frontline personnel. Aaron shuddered as he contemplated all that could go wrong, as things had so often in the past. Some hands went up to ask questions and to provide the inevitable kind of pushback which had sent all the other improvement initiatives off the rails. The Doc fielded the first inflammatory question, then shut down further flak saying, "You are here to learn and apply, not to debate and obfuscate. If you do not want to assist this organization to improve, it is time for you to exit and give your position to those who will help deliver the dramatic turnaround your organization needs if you are to stop injuring and killing your fellow employees." The CEO smiled, nodded his head, and the training began.

The Doc's first story discussed his involvement in assisting a heavy manufacturing organization which was comfortable using Total Quality Manufacturing (TQM) and Total Productive Maintenance (TPM) techniques. They had some small continuous improvement (CI) teams which engaged in solving the frontline day-to-day difficulties that commonly occur in the day-to-day operations of organizations worldwide. The people of this organization felt they were pretty good at what they did and questioned the need for another CI process, especially one that took dedicated team efforts over a relatively long term.

To them, and to Aaron's group, the Doc described the purpose of the Continuous Excellence Performance (CEP) process. CEP is an engagement means for resolving safety issues by empowering teams of employees in a structured problem-solving process. The Doc discussed why a classical frontline CI team concept often struggles with delivering in-depth solutions to the more complex safety problems. He explained that day-to-day problems are like swatting the flies which buzz around

Delivering Safety Excellence: Engagement Culture at Every Level, First Edition. Michael M. Williamsen.
© 2021 John Wiley & Sons, Inc. Published 2021 by John Wiley & Sons, Inc.

and need to be eliminated. A small frontline employee group comes together and, relatively rapidly, solves the equipment-related issues which typically occur with TQM and TPM. The Doc's judgments were that TPM type maintenance teams dealt with about 90% equipment-related issues and about 10% human interaction. The TQM approach, within the quality environment, was felt to be more like 70–80% equipment/product-related process issues and 20–30% human interaction.

Aaron's group took the opportunity to respond and agreed that in safety, solving the day-to-day reactive Level 1 and Level 2 equipment and condition issues (explained in Chapter 10) occurs relatively quickly with a fairly classic CI team approach. However, once the hardware weaknesses are mostly in control, the focus quickly shifts to about 90% people/behavior realities, which cause the overwhelming majority of workplace incidents/injuries. As the Doc was about to launch into deeper training one of the audience members asked where all his charts and figures came from. In answer, the Doc replied that most of the materials in the teaching they were being presented was developed over time by the Doc, his associates and the various customers who had engaged in the CEP process worldwide.

Herein is the major issue for the frontline CI team approach. In safety, culture and behavior process issues require far more time and involvement intensity than is available with the typical one- to two-hour CI team event. These more in-depth safety process Rapid Improvement Workshop (RIW) teams are monitored and their progress is adjusted by a cross functional, cross organizational Safety Steering Team. The RIW teams are made up of hourly, supervision and upper management personnel who are trained to be involved with the more intensive error proofing of upstream human processes. This entire procedure results in error proofed processes with well thought-out and tested people components, all of which help to deliver sustainable, downstream, low safety incident number realities (Figure 18.1).

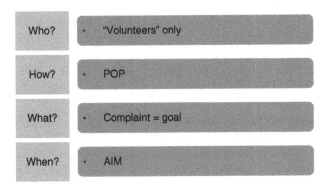

Figure 18.1 Four Ws for forming a team.

The RIW cross functional teams typically meet off and on over a 60–90 day time frame. During this time frame they develop and then actually do test runs to prove a complete solution to the safety process in question. Their in-depth solutions include components like: pilot trials, accountabilities that help deliver solutions, practical accountability-based audits, policy statements allied with the processes, "cookbooks" (instructions) as to how the total organization delivers the interpersonal results, etc. This approach is way beyond what a one- to two-hour CI Team can accomplish. Likewise, the upstream and downstream results are noticeably beyond that available via day-to-day fly swatting. However, both approaches are necessary to get a zero incident/at risk activities safety culture to occur and to be sustainable in the long term.

18.1 Rapid Improvement Workshop Teams

The real work is done by the Rapid Improvement Workshop (RIW) teams that the safety steering team oversees. The RIW teams are made up of volunteers who come from across the organization and from all of its levels. On the first day of the workshop, the 8–12 members of each team learn the differences between traditional approaches to safety management and those which deliver world-class results. They discover along the way that successful organizations manage safety the same way they manage every other important business initiatives. Over the next three days, the team is guided through an intense engagement journey to the final solution by a safety professional who is a trained facilitator in the RIW process. The team agrees on (1) the purpose, (2) the deliverables, and (3) the approach which will be used to achieve a solution. This is the team's POP (Purpose Outcomes Process). They are then trained to use a set of simple, effective solution tools which include process flow diagrams, Pareto charts, cause and effect diagrams, etc. that are common tools in CI endeavors. They also learn how to turn problems into goals along with a number of other practical team engagement and solution delivering techniques (Figure 18.2).

At the end of the three- to four-day workshop, the team has an outline of a viable solution for the issues they have been asked to resolve. Additionally, they have an action item matrix (AIM) which lists all the remaining tasks, indicates who will work on the tasks, and includes a timeline for accomplishing each line item. The entire plan is scheduled to be achieved within a 90-day time period. Armed with a solution and the personnel and means to deliver it, the RIW team presents its solution to the Safety Steering Team at a two-hour briefing on the final day of the workshop. The Safety Steering Team reviews the plan, and gives feedback as to scope, timing, personnel engagement, and viability.

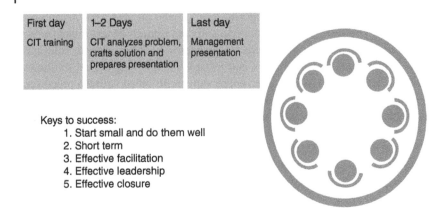

First day	1–2 Days	Last day
CIT training	CIT analyzes problem, crafts solution and prepares presentation	Management presentation

Keys to success:
1. Start small and do them well
2. Short term
3. Effective facilitation
4. Effective leadership
5. Effective closure

Figure 18.2 Rapid improvement workshop. Source: © Williamsen Enterprises.

There are a number of typical items RIWs focus on: near miss/close call Resolution (NMR), on boarding, aging employees, training, disaster time, effective safety meetings and tailboards (work site pre job safety discussions for the tasks, equipment, and conditions at hand), recognition, hiring outside contractors, etc. These are just some of the many deeper issues that can help an organization achieve safety culture excellence, and the ongoing organizational and personal discipline necessary for this new safety culture to be sustainable.

The Doc has seen and worked through many, if not all, of these usual suspect difficulties before. So Doc, why do not you or your counterparts just give the RIW team the answer, speed up the process, and reduce the customer's cost?

In answer to this challenge the Doc commented that there are a number of safety cultures within every organization. They have required processes to manage:

- The government regulations which focus on conditions, policies, and procedures.
- The culture which reacts to what is seen (both corrective and positive reinforcement) with respect to safety in the workplace and off the job.
- The safety accountabilities which come in to play for hourly, supervision and management personnel both on and off the job.
- The engagement of people from all walks in your organization who must focus on improving what just is not good enough for all those things listed above.

In the ever changing organizational dynamic there is a need for all involved to learn how to engage in making the workplace, home, and recreation safety cultures those that just do not tolerate what can lead to injuries. The dynamic complexity does not allow an organization to be thrown a "fish" when it comes to getting better. The answer to all of this depth and clutter is that people of an organization need to learn "how to fish" so they can solve the safety problems which inevitably exist in both their working and off the job activities. Tossing them a canned set of answers does not develop the strong effective culture that is necessary for consistently delivering excellence in all that needs to be done over time.

To further illustrate this point, the Doc shared a story about the famous psychologist Ivan Pavlov. Pavlov showed it was possible to train his dogs to salivate, but he could not educate them: they had no real knowledge. How much of your training is classical conditioning? Is it "salivational" and not the retention and utilization of value-added knowledge?

A viable organization needs to challenge its people to get them to think so they can respond correctly when they are outside of the classical conditions. They have to be able to think and analyze unique situations and then propose reasonable solutions to the realities of the challenges they face in the world they live in.

A case in point deals with check lists. All organizations have them, but do the check lists cause people to think, or just apply a "check in the box" type of salivation? Well-designed and detailed checklists can cause people to think/consider items beyond their normal exposure. The Doc then commented that he had learned the value of good complete checklists from a technician who had previously served in the military and was familiar with the military checklists. He developed good checklists which caused workers to identify both problems and possible solutions to complex systems. An exercise for your teams that have checklists is for the team to develop their own improved checklist (or modify a good existing checklist as needed) based on their own experiences in the process. Good planning and creativity are required to develop a good/effective checklist that not only covers the basics but also requires creative input. As you work through the many issues needing improvement in your organization consider developing a process of "continued improvement" for the checklist(s). The ones you develop for system evaluations not only check the boxes but also require the evaluator to provide "creative and objective evaluations" needed for improved system operation. Additionally, how can you get your checklist and its information to be of value up the ladder of the organization?

This RIW learning process is a deep dive into knowledge which can be effectively used on and off the job (Figure 18.3).

Fortunately, the necessary tools are not complex and do not require complicated math – rather they are like basic tools in a toolbox that are used to work on complex machines. They are creative problem solving tools. The RIW teams are taught how and when to use these simple, but powerful tools/concepts like:

- POP (Purpose Outcomes Process)
- AIM (Action Item Matrix)
- Complaint Equals Goal

❑ POP statement ❑ Pareto chart
❑ Complaint = Goal ❑ Pareto voting
❑ Cause and effect diagram ❑ Process flow chart
❑ Five WHYs ❑ Action item matrix
❑ Affinity diagram

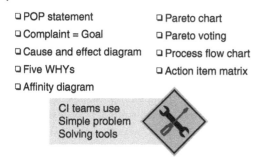

CI teams use Simple problem Solving tools

Figure 18.3 Tools used by CI teams.

- Cause and Effect Diagrams
- Fault Tree diagrams
- Process Maps
- And the like.

Learning how to use these simple problem-solving and group engagement tools can revolutionize how an organization becomes much better and much faster at solving issues with respect to: safety, productivity, quality, and customer service.

The bottom line is that an organization needs to develop a culture which solves its own problems. An organization can only do this by going on the journey with some of their skin in the game.

With a plan agreed upon by the SST, the RIW team meets regularly to deliver a final solution which includes:

- Necessary training
- Safety accountabilities across the organization
- Communication
- Auditing.

These and other needs delivered by the RIW are required to achieve a practical, workable solution. Each month an RIW representative presents the current solution process progress to the Safety Steering Team for guidance and feedback. The ongoing task completion and progress review continues for up to 90 days. The proposed solution then goes into a pilot phase in which it is implemented by a target work group that shakes out any remaining details. The pilot progress is also reviewed monthly by the Safety Steering Team, as is the ensuing solution dynamic for rollout to the entire corporation. The team's detailed documentation is important. It needs to be both understandable and usable in the organization. As a result an effective scribe is a valuable team member.

Practical resource realities limit an organization to annually do about one to three significant RIW team solutions with their associated pilots and rollouts. This ongoing solution development, piloting, rollout, and field viability monitoring keeps the Safety Steering Team busy. In so doing, the SST members stay focused on proactively delivering needed solutions to safety issues, rather than just reacting to injuries.

18.2   Delivering a Better Safety Performance

Achieving cultural transformation does require an investment in both time and money. Improvement requires both financial and personnel resource commitments to change what an organization is doing. However, time and again this approach has rapidly and effectively delivered safety performance beyond classical safety approaches merely focused on reacting to injuries and events.

At this point the Doc shared a personal story about his interaction with a large global petroleum company on the North Slope of Alaska. When he presented the safety perception survey results, the vice president (VP) of drilling immediately challenged one of the three lowest scoring safety leadership processes (which was substance abuse). The VP went on about their one strike and out drug policy, their random drug testing process, and their significant focus on this issue. Surely the survey data must be wrong because their culture of limiting substance abuse was very strong! Before the Doc could reply, one of the drilling contractors stood up and shared their firing of a crane operator a month earlier. When they examined the former employee's locker, they found 1000 oxycontin pills. They then analyzed the random drug sampling policy the large petroleum company had them utilize. The contractor's group determined the field forces had broken the code of when test (urine) samples would be collected. The under the influence people were able to wear urine clean bags so they could pass the test on the appropriate "random" test days. The whole demeanor of the meeting immediately changed from confrontational to participative. About a year later the petroleum company sent a note to the Doc, thanking his group for the data, insights, and process it provided for their company to significantly improve their North Slope safety culture.

Organizations have found that investing in this focused Safety Steering Team and Rapid Improvement Workshop approach has helped them to receive a definite positive return on investment (ROI) which includes:

- Fewer injuries with their associated direct and indirect expenses.
- Multiple knowledgeable safety champions at all levels of the organization.
- Improved working relationships and morale between hourly and salaried employees.
- Elimination of high and low risk issues which can lead to injuries.
- A culture of safety accountabilities which proactively eliminate the possibility of injuries.

The bottom line is that the Safety Steering Team and RIW approach delivers a safety culture which continually error proofs safety processes, conditions, and activities in the relentless pursuit of a zero incident/at risk activities safety culture, both on and off the job.

All organizations have training, background experiences, desire, abilities, duties, DNA, etc. In short, each has its own culture, which is made up of a complex formula unique to their selves. Their employees are then called on to use this personal, private mass of talent on their job, at home, on vacation, and elsewhere in the lives they lead. This call to action can occur at a moment's notice, or as a part of a carefully contrived plan which takes significant time to develop and execute. Sometimes, the call to duty is a significant use of each of the employee's abilities and makeup plus the group makeup of many employees. Sometimes, there is a unique nature in what is to be accomplished. And thus the rub: what happens when people are expected to break out of their traditional approach, or comfort zone, and do something new? Can they get out of Popeye's cartoon character syndrome: "I am what I am and that is all that I am"? If not, then they only regurgitate the same old safety chime that has been digesting forever, and in reality, is just passing through their system.

Are organizations only capable of delivering the same old: videos, check sheets, policies and procedures, observation system, regulations-based condition fixes, ad nausea? Is the solution approach really just the classic definition of insanity? As safety performance gets better and better there is a need for new approaches which can help an organization break through the performance plateau that the same ol', same ol' culture cannot solve.

If an organization wants to achieve a performance level beyond the no longer acceptable results delivered by the classic tools, it must go beyond regurgitating and start engaging the people across their organization. As a culture is faced with a plate of problems, its leadership must consider how to engage others in using their talent mass to deliver creative solutions, which have the capability to clean up what remains "spit up" in front of us. The "spinach of the past" has but the power of vomit in solving the many unique issues which are faced by today's safety professionals.

> *At the session end, Aaron was impressed with the team's attention and engagement, and only one person dropped out. After a lunch break, the next training topic was going to be a real challenge: safety accountabilities. In all Aaron's time with this company there had never been a successful accountability initiative. The felines resisted herding, no matter what incentives or punishments were tried. Aaron hoped the Doc was having a power lunch, because he was certainly going to need it.*

19

Fundamentals That Are a Result of Developing a Culture of Safety Excellence

Aaron sees the Doc returning from a lunch with his CEO and wishes he could have been the proverbial fly on the wall to listen in on their conversation. He wonders about the safety accountability section next up on the agenda. What will his safety accountabilities be? What about those for upper management, middle management, frontline supervision, frontline union employees? For some understandable reason, the theme music for Mission Impossible is play- ing in Aaron's mind. The stage is set for "do or die," "make or break," "put up or shut up," "the bottom line" and a host of other clichés that say "the moment of truth." Aaron has a significant amount of angst weighing heavily upon him.

The Doc starts off with a story about a proud, but struggling company on the Pacific Rim. He was given an in-depth tour of their manufacturing facility. The organiza- tion was a global leader in their high technology product which also had highly toxic components. At the end of the extensive tour, the Doc asked his Asian coun- terpart to bring together hourly and salaried personnel who wanted to help their company improve. To the Doc's consternation, only salaried personnel showed up. When he complained to his counterpart, the Doc was told these people were very smart and hourly personnel were not really needed to work on their specific issues.

There were some definite statistical issues in their processes and thus the Doc asked them about their understanding of Taguchi[1] concepts and approaches. He was assured the managers were very competent in this approach. Over the next hour the Doc pursued the issues with intensity. At the break, his coun- terpart pulled the Doc aside and apologized, saying he was "embarrassed that his managers didn't know s@##." He would have hourly people in the next session after lunch. When the break was over, the room had management on one side and hourly on the other. Neither side communicated with the other. Such communication was not part of their culture. People could be embarrassed if they were confronted with their lack of knowledge on manufacturing issues.

Delivering Safety Excellence: Engagement Culture at Every Level, First Edition. Michael M. Williamsen.
© 2021 John Wiley & Sons, Inc. Published 2021 by John Wiley & Sons, Inc.

The Doc performed some serious ice breaking exercises through the translator. Part of the commentary dealt with the fact that organizations seldom have a lack of talent. There's plenty of intelligence to solve the problems, so intelligence is not the issue; it's the realities of a weak culture that keeps an organization from engaging all the available brain power. Two days later, as the Doc headed home, this important Pacific Rim supplier had broken down barriers which had existed forever. They were using all their people's abilities, no matter what level of the company they came from, when they were needed to participate in solving problems.

The point of this account is that change is often a difficult cultural challenge. Improvement is difficult. If a struggling company is to emerge victorious, the people at all levels must seriously work together on the issues which are causing them to stumble and fall. The people across the whole organization must break the old paradigms and barriers and work as a team that engages all the skills and brain power they can bring to bear.

With that said, in the change/improvement process, the five-step accountability model and how this is a deliverable of the RIWs become a mission critical part of developing a culture of safety excellence.

"IT'S NOT MY JOB!!" is one of those excuses which drives leaders up a tree. As he was growing up "accountability" was a bad word to the Doc. It meant he was being held responsible for something else that could easily go wrong, or to change what he wanted to do to what someone else wanted him to do. Along the way, the Doc found many, many people who felt the same way about this nasty word "accountability." However, if people look at how businesses and organizations function effectively, they are all about regular activities (accountabilities) that people are expected to do well, and on time. Therefore, the Doc needed to redefine accountability, both for himself and for the myriads of people in the many organizations within the global world of work. The definition the Doc now uses for accountability is: "You can count on me to be able to do this job safely and correctly ALL the time." As people take on new tasks, or changes in task content, or a technology upgrade that requires additional training and proof of competency, consequential changes must also come into play in order to keep personnel safe, correct and current in the work being performed. "IT'S NOT MY JOB!!" just is not a realistic option for the masses of people, the Doc included, who must deal with the realities they encounter in their lives – both in the workplace and beyond.

Accountability is perhaps one of the most talked about concepts in business today, yet one of the least understood. Leaders typically view it as something they must hold others to, while employees see it primarily as discipline for something they did wrong. These widely held perceptions reveal only limited aspects of a strong accountability system. There is much more that leaders across organizations must understand, and apply, in order to build the voluntary and accurate execution of workplace accountability.

Accountability is a deep topic that can lead to long discussions and a level of intensity. One definition does not usually suffice when working with groups. As the Doc has worked with many organizations each seems to develop their own definition that fits who they are. Even the five-step chart below is a modification of the three steps Dr. Petersen originally envisioned. As improvement teams work through their issues they too modify work group accountabilities to fit local reality. Another way to view accountability is that of accepting responsibility for, and providing satisfactory execution of, one's own actions and deeds. It is the opposite of blaming others for things that go wrong. Leaders can either *hold* their subordinates accountable for the expected work, or they can create an environment in which others *take* accountability for their work. When employees take accountability for their work, they do what is expected of them with a high degree of accuracy and in a safe manner while requiring very little intervention from the leader. This process as improved by the Doc's associate David Crouch requires a one-on-one relationship between leaders and subordinates (Figure 19.1).

Safety accountability: Safety accountabilities are the visible, well-defined activities completed by an organization's personnel that are designed to control/eliminate risks. When safety accountabilities are clearly defined, and actively practiced at each level of the organization, a culture of safety excellence can be created.

Strong accountability systems include five steps for each level of the organization:

1. **Clearly define expectations:** what activities must be done for safety.
2. **Train to ensure competence:** does the individual have the ability to do the job safely in a quality manner?

Figure 19.1 Accountability model.

3. **Provide necessary resources:** what is needed to accomplish these accountabilities.
4. **Measure the accuracy of execution:** the required quantity and quality of these efforts.
5. **Deliver appropriate feedback:** at task completion to motivate further desired behavior.

When properly designed, accountability systems are the engines that power and enable a culture of Continuous Excellence Performance (CEP). Keep in mind that accountability is a noun, and being accountable is a verb.

Safety accountability realities: Accountabilities which help deliver good safety performance are hot safety topics of our times. Many people have been a part of teams who have studied and delivered safety accountabilities for various levels and work groups within their organizations, and then their safety accountability documents are filed away in safety's equivalent of the dead letter office. When this happens, safety accountability fails, it just became a noun instead of a verb.

There are a number of reasons for this demoralizing reality:

- Do safety accountabilities really get measured and significantly rewarded/recognized?
- Has upper management been involved in developing the safety accountabilities they and others will have to perform?
- Has the five-step model of accountability been followed: Define, Train, Resource, Measure, Recognize?

The real shortfalls for safety accountabilities are resolvable. To become a productive initiative, both your team and your people's follow-up efforts will have to focus on the practical realities of transitioning safety accountability from "accountability" to "becoming accountable."

There is an involvement process which must be followed in order to be successful with safety accountabilities, and "Define" is only the first step.

One of the many common problems in work site safety deals with the metrics of engagement and performance across the levels of an organization. In the chart shown below:

- **Results** typically measured by upper management deal with injury statistics, indicators of what we do not want to happen. This indicator lets upper management know how their organization is doing numerically, but gives no data on what needs to be done to improve the safety results. It is just a number, and that number has no real value or usefulness which creates action at the frontline where most all the incidents take place. At the frontline, results numbers like RIF are almost of no value. Since injury statistics are calculated on 100 effort years (100 years of work), and the average supervisor has a crew of less than 15 people, one injury would take about 6–7 years to disappear from their record.

- **Activities and accountabilities** directly impact safety performance. Are the people throughout the organization regularly doing proactive tasks which reduce the possibility of an injury or incident? Such practical, effective safety accountabilities make a huge difference in safety performance. However, they are seldom a part of the middle management or executive compensable metrics. This upper level workforce needs to have both active and value-added engagement in the safety processes, with measurable, value-added accountability metrics. When looking at the chart it is obvious that most safety accountabilities occur at the frontline where the most risks are encountered. Nonetheless, for middle and upper management daily or regular value-added safety accountabilities must also be a part of their safety culture.

- **Effective safety systems** have a well thought-out and proven set of diagnostics and engagement activities which deliver visible solution focused accountabilities at each level of the organization. This kind of engagement across the organization makes a positive difference in the safety culture and its performance of excellence. There are no "IT'S NOT MY JOB!!" excuses at any level when well thought-out and proven accountabilities are just a part of what is done every day in a correctly functioning safety culture (Figure 19.2).

One of the critical deliverables of the RIW solutions is a set of realistic, practical, effective safety accountabilities for each of the appropriate responsible parties in the process being improved. These then become the measureable metrics in the five-step process of accountability.

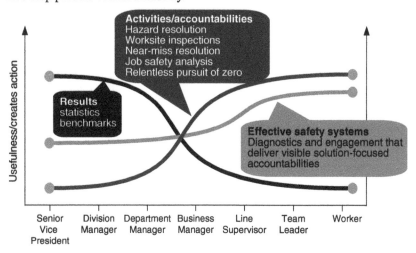

Figure 19.2 Accountability measurement model.

Aaron leans back and observes as the next break has people talking about sensible safety accountabilities for themselves and others. He smiles and wonders if the leadership is finally beginning to learn their "s@##." Aaron looks forward with some trepidation to working with the teams in this whole new concept of daily/regular activities, practiced by all levels of the organization to help eliminate incidents.

Note

1 Genichi Taguchi was an engineer and statistician. From the 1950s onward, Taguchi developed a methodology for applying statistics to improve the quality of manufactured goods.

20

Communication and Recognition

Aaron wonders what is next? The team is saturated with input. They need time to absorb all the content. After some time of Q&A, the Doc calls it quits for the day. He then asks the team to consider what kind of feedback, or input, would help reinforce the importance of people performing their individual safety accountabilities? What is recognition, how does it fit with accountability, how is it communicated? This leads to the very important factor: How is it used?

Aaron goes into mild catatonic shock. His autocratic upper management has typically used only punishment as a motivator. Once again, the Doc is about to enter the twilight zone of uncharted resistance to change. Aaron sure hopes and prays that this big hole in their weak culture can be filled. But nobody can bat a 1000 . . . . Can he?

The Doc kicks off the next day with some more of Dr. Dan Petersen's famous words for struggling safety cultures: "What gets measured is what gets done. And what gets rewarded is what gets done first." The traditional accountability system shown in Chapter 14 puts the recognition focus on what we do not want to happen, injuries. This is not a significant driver for performance excellence: measuring and rewarding (recognizing) the quantity and quality of daily activities which reduce the possibility of incidents. The injury targets set by organizations, such as those Aaron lives in, typically get blown out of the water with the occurrence of one or more injuries. And then people give up and just live with "The beatings will stop when the safety performance improves."

OSHA, VPP, classical safety teaching materials do not include much, if any, upstream noncompliance-related activities. Practicing, measuring, and recognizing high-quality daily and regular safety accountabilities are what is missing in the typical government regulations (regs) model. This kind of safety accountability

Delivering Safety Excellence: Engagement Culture at Every Level, First Edition. Michael M. Williamsen.
© 2021 John Wiley & Sons, Inc. Published 2021 by John Wiley & Sons, Inc.

approach definitely helps build a strong interactive safety culture with its resultant excellent performance.

The Doc mentions one of his favorite books on recognition, *The One Minute Manager*, written by Dr. Kenneth Blanchard.[1] In the storyline of this book a formerly ineffective leader learns, and then practices the basics of improving human interaction:

- Make your communication about personnel performance issues (either good or bad) both crisp and of short duration: about one minute.
- Look for what the employee is doing correctly and reinforce their correct work.
- Try to deliver an abundance of positive feedback: about seven times as much positive reinforcement as negative correction (coaching).

Aaron chimed in that he had read the book and commented that he found it a real struggle. Aaron thought two to one positive was way more than he ever considered possible. "Seven to one, you gotta be kidding!" Aaron's classical regs safety training was that safety people always look for what is wrong and then confront errors and sloppy practices. To which, the Doc's input was: "We do have to address the wrongs. However, if you really think about your workplace realities you will conclude, as I did, there is a whole lot more of what is right going on than what is wrong. Indeed, people are paid to do what is right. However, to get better performance your emphasis needs to be on positive reinforcement of the many things people are doing correctly. This is another coaching model."

Leaders must provide timely, relevant, specific, and frequent feedback to subordinates about the work they are expected to perform. *Positive recognition is much more effective than correction or criticism.* Most people work safely, most of the time, yet the majority of the communication they receive from their leader is about the unsafe work they do. This is a problem. The basic principle is this: the more you recognize the safe work your people do, the less you will have to deal with their unsafe work. Positive recognition is a very powerful tool used to influence the performance of others. When a leader builds it into his/her own habitual leadership behavior, others will gradually begin to self-correct the things they do and reinforce the positive recognition approach for improving the safety culture.

Despite some organizations' best intentions to incorporate a form of recognizing good safety performance, it often rates among the lowest safety culture processes (categories) in safety perception survey results. Too often, recognition strategy is confused with awards/rewards, such as certificates, contests, cash, and trinkets. Excellent safety recognition is all about establishing and using an approach to effectively communicate with workers when they are doing a good job. A strong recognition system refers to the one-on-one interpersonal interactions which focus on what an employee does well day to day.

20.1   Encouraging Positive Behavior

Behavior modification is nothing new. The basic process involves systematically reinforcing positive behavior, while at the appropriate times using negative reinforcements to eliminate unwanted and unsafe behaviors. There are two primary approaches used in behavior modification programs: one is an attempt to eliminate unwanted behavior which detracts from attaining an organizational goal; the other is the learning of new responses.

In safety, a primary objective is to eliminate unsafe acts. This approach leads to behavior modifications which create acceptable new responses to an environmental stimulus, including an unsafe behavior. The basic concept is a systematic or consistent way to carry out positive reinforcement. Its result leads to improved performance in the area to which the positive stimulus is connected. This concept is based on the simple formula: $B = f(C)$, which means that a person's Behavior (B) is a function of (f) the Consequences of past behavior (C). If a person does something, and immediately following the act something pleasurable happens, he or she will be more likely to repeat that act. If a person does something and the act is followed by something painful or unpleasant, he or she is less likely to repeat the act (or at least is unwilling to get caught next time).

Positive reinforcement is essential in today's work environment. Numerous studies show positive reinforcement increases worker productivity and quality, improves labor relations, reduces absenteeism, and bolsters employee retention. It works the same way when encouraging safe behaviors.

Safe behavior reinforcement is nothing more than recognizing people when they do a good job, at the time they are doing it. Recognition is often the most underutilized management technique, and yet it is among the most powerful.

Each of the safety culture processes (categories) in the safety perception survey has a fault tree diagram. The fault tree diagrams help guide the teams in analyzing and working through the important aspects and details of each safety culture process. The fault tree diagram shown below is an excellent tool for developing a strong recognition process that supports effective safety accountabilities.[2]

The resultant training for recognition emphasizes the need for strong one-on-one communication. This training uses a simple approach to reinforce the need for good recognition communication. The end result is training people in how to effectively recognize truly good performance. Their one-on-one communication must be:

- Timely: "You see it, you say it."
- Relevant: Specific to the action or behavior you want reinforced. Make it personal.
- Sincere: Do not patronize, or belittle.

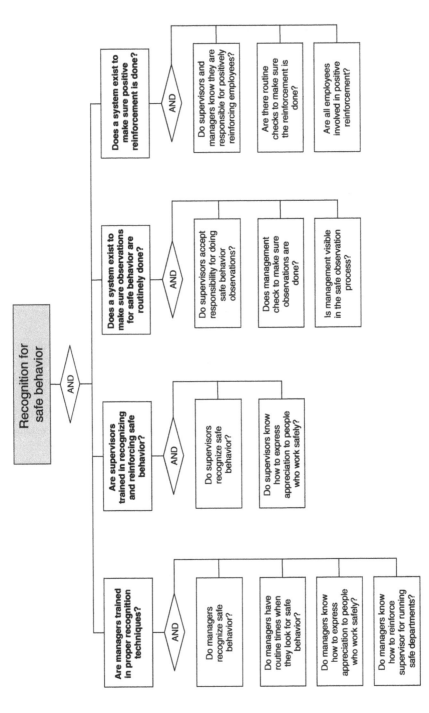

Figure 20.1 Safe behavior recognition.

- Confirmed: At least have eye contact and between the ears head agreement. Ask open-ended questions and verify that the person understands why they are being recognized.
- Frequent: Consistent with no favoritism, without overdoing it (Figure 20.1).

> *Besides the recognition flow chart, the Doc showed two short videos about effective feedback techniques; "Speak Up-Listen Up" and "Recognize It."[3] The material provides a significant amount of detail on how to develop and deliver this kind of proactive safety culture of correct. Together, the videos removed any significant uncertainty about positive reinforcement and how an organization's employees can communicate and reinforce what is correct and what needs to improve, without resorting to "psycho-babble." Such a culture shift seemed very possible, reasonable and desirable. Once again, Aaron hears and observes significant discussion from the other team members. The entire dynamic of Aaron's organization was changing in front of his eyes. Maybe there is a light at the end of the tunnel, but there is still more required to accomplish the desired result.*

Notes

1 William Morrow and Company Publishers, 1982. ISBN: 9780062389152.
2 Original publisher CoreMedia Training Solutions in a booklet titled The Challenge of Change.
3 Original publisher CoreMedia Training Solutions.

21

Hazard Recognition Is Different than Hazard Control

> *Aaron looked at the agenda and saw that next up on the agenda was a segment describing huge stumbling blocks for his company: hazard recognition. The managers, who spent very little time in the field, probably wouldn't recognize most of the day to day hazards of the work place even if they were bitten by one. The frontline employees seemed oblivious to hazards; after all, they get paid to work around this kind of stuff every day. Adding the managerial hazard recognition "zero performance" to the frontline hazard recognition "zero performance" added up to their company being the 24th worst RIF amongst the 25 largest companies in their industry. Aaron thought all this safety culture excellence material added to the OSHA regulations and compliance materials was supposed to solve all the hazard issues. "What gives, Doc?"*

The Doc kicked off the session with a story about a team working at an underground coal mine in America. They were impressed with the safety protocol which required a methane (deadly coal off gas) sampling/alarm device at the mine workface. In addition, there were water sprays to keep the coal dust from being breathed by the miners, which could lead to the infamous black lung disease which has caused the deaths of so many miners. As the team went underground they were mortified to find the water sprays directed at the methane monitors so that these nuisance alarms would not shut down mining operations and cause the workers to miss the $0.50/hour bonus for making productivity standards, aaarrrrgghhh!

The Doc introduced the work of Dave Fennell, a good friend and a most excellent safety leader. Dave was the senior safety advisor for many years with the Imperial Oil division of ExxonMobil Company in Canada. Throughout his tenure there, Dave was frustrated by all the hazard recognition training, which seemed to make little or no real difference in injury rates for the field workforce. After more than two years researching this issue, Dave changed the whole approach of hazard recognition to include what the real solution appeared to be: **Personal Risk Tolerance**.

Delivering Safety Excellence: Engagement Culture at Every Level, First Edition. Michael M. Williamsen.
© 2021 John Wiley & Sons, Inc. Published 2021 by John Wiley & Sons, Inc.

21.1 The Common Threads

In the workplace, at the home, and off the job, everyone makes untold numbers of decisions about what to do and how to do it. From a personal risk perspective, there appear to be three common threads:

- **Recognizing the Hazard.** The uninitiated worker may not actually know what the hazards are in the workplace or know how they could personally be impacted by those hazards. Most parents can remember circumstances experienced in raising their children when they had to step in to protect them from dangers the children honestly did not know about, or appreciate. For a similar reason, new employee orientation is of significant importance to the newbies who just do not know what they do not know. There are all kinds of statistics available about a high percentage of injuries occurring to the newer employee who is not aware of existing dangers, or the mitigating factors he/she should employ. The use of good/thorough safety training and a supporting Job Safety Analysis (JSA) process is important for recognizing the hazards which will be encountered with work tasks. The JSA must go beyond just recognizing the hazards and into what must be done to eliminate the possibility of injuries. Additionally, there must be a corporate and personal commitment to both train and live the JSA realities.
- **Understanding why it is a risk**. Employees all come with differing sets of perceptions based on things like age, physical capabilities, pride, previous experiences, and the like. These personal filters are often very effective in keeping people from correctly evaluating the problems commonly faced. In turn, this kind of false confidence carries with it a noticeable probability of resultant personal injury. Three simple questions can help a worker understand the risk: "How could this impact me?" "How bad could it be?" and "What are the significant factors involved in creating a hazard?"
- **Risk tolerance.** When people see the hazard, and understand the risk, do they make a conscious decision to remove/reduce the hazard, or do they just proceed with the work? At this juncture, the Doc showed one of Mike Rowe's "Most Dangerous Catch" short TED videos.[1] The Doc then shared Rowe's perspective: our safety is not dependent on what the government or the company provides us. With respect to safety, our safety, first and foremost, depends on an individual's personal common sense and the personal responsibilities/precautions they take when faced with issues that can be dangerous to them, their fellow workers, and their family members.

The Doc then presented focus material on the 10 most frequent personal risk influencing factors. The Doc mentioned that Dave comments to his audiences

---

1 1 https://www.youtube.com/watch?v=XuvyS63iK34 (Accessed 21 June 2012).

these are basic human characteristics which we need to understand, rather than faults which need to be corrected. This personal comment seems to generate more buy-in when framed as follows: why people normally accept certain levels of risk, and what you and your employees can do about this personal risk assessment reality, both on and off the job. This personal risk assessment concept is a significant shift required for improving industrial safety. No longer are the limits focused on the available tools which exist with government requirements, observation programs, and CI initiatives. Personal risk assessment takes the actions, responsibilities, and necessary thought processes for safety to the employees at all levels of an organization. The other tools are an important part of safety programs. However, when practiced and lived, personal risk assessment ties all these tools together while adding the principle of each person having the need to analyze, focus, and execute on his/her own surroundings, capabilities, strengths, weaknesses, and attitudes. If safety is to be, it will be up to ME to make it so. "I am responsible for my own safety."

Following are the 10 personal risk assessment realities which must be considered by ourselves, our employees, and our family members.

21.2 Overestimating Personal Capabilities

This is the first of the 10 common personal risk assessment dangers. The Doc's personal story happened way back when he joined the military and went to boot camp. "Young and stupid," physical fitness at its peak, "10 feet tall and bullet proof" – all these and other attributes, like adrenaline and commitment, made each of the raw recruits confident they could take on the world and come out unscathed. It did not take very long for the Doc to realize he needed far more than personal confidence and physical fitness to come out of the Marine Corps unscathed. Just having confidence in his abilities would not keep him or his team members safe and uninjured in the real-world dangers of the military experience. Common sense, personal responsibility, team work, training, and attention to situational reality were absolute must-haves. It was drilled into "the boots" daily.

As the new recruits enter an organization, they encounter unfamiliar dangers. They are likely to be overconfident in their personal abilities, and to underestimate the dangers of the job at hand. Like many new employees, they are at high risk, and try very hard to prove their worth on the job with a "get 'er done" attitude. The young and vigorous can lift more and work longer in challenging environments, and they often feel a personal desire to prove themselves to others plus numerous other factors. When it comes time to correct this behavior, challenging a worker's ego and pride does not seem to work very well.

There is a second class of employees who also suffer from overconfidence in their capabilities: the older, experienced hands who know the job and the dangers, and they still take shortcuts. They believe they can achieve without injury because of their personal experience factor. After all, they have been on the job for years and this is just the way their work group operates, it is their norm. And norms are very hard to change. So is the personal risk assessment culture in which they live. They just do not seem to be aware of, or comprehend, the dangers they ignored and lived with all those years.

The Doc mentioned that his Papa had a saying that went something like, "Change only occurs when the pain of no change exceeds the pain of change." The Doc then gave an example from his work with an electrical utility company. The previous week, a contractor had two of its employees free climbing a tower like they always did, instead of using the slower required fall protection. The upper employee fell and took his friend down with him; they fell 120 ft to their deaths. During a training break, two linemen described how each of them had done a similar free climb and fell about 40 ft in separate events. Obviously and amazingly, both survived, but more than a decade later both still suffer from pains associated with their injuries. Both were able to go back to climbing towers, but the pain of no change had them always using fall protection equipment, and taking the time to do so.

How to handle overestimating personal capabilities? The "young studs" often do not handle personal challenges from superiors well at all. The seasoned hands often do not believe they will ever "roll snake eyes." Both seem to charge blindly on, in hopes that they will never walk off the edge of the cliff they know is out there.

In military cultures, there are rules that MUST be followed. In safety, these rules of engagement are often called JSAs (Job Safety Analysis). JSAs must be well written and have the practical field content which gives them credibility with the people who are performing the tasks. These JSAs must be included in training and practiced by the field hands doing the job every day, for every job to which the JSAs apply. Supervision and upper management must consistently, visibly, and frequently reinforce the importance of following these rules of engagement. There must be a "culture of correct" which is lived every day on the job so no one is put at risk of dying on the job any day. It goes something like; "I don't care if you are in such good shape you can lift 200 pounds. For this company, no matter where you are working, you will get assistance lifting anything weighing more than 50 pounds." There are similar rules of engagement in this kind of "culture of correct" for each personal risk that the job at hand may present.

One of the struggles of living this culture of correct deals with how hourly and salaried leadership consistently provides verbal feedback, i.e. how they effectively deliver the message. Another policy or procedure is not the answer. Physical presence and adult conversations do a much better job. This is an important aspect of all 10 of the personal risk factors.

21.3 Complacency – Familiarity with the Task

Repetitive tasks exist in most workplaces. Wherever they are present, they are also a frustrating safety risk. The more we perform a particular task, the better we should be at it. This is true for the first few times a person does a task, but then human minds have a tendency to wander. The safety principle of eyes on task and mind on task becomes cloudy, and to some extent, people become complacent. Each and every person has a different complacency threshold, whether it is the 50th or the 100th time. People get sloppy, the risk of personal injury increases, and the probability for personal injury becomes ever greater.

When experienced hands become complacent and unfocused, there is a tendency to pass this dangerous autopilot mindset on to new hands, furthering the problem. One of the better safety pros suggested some simple but effective techniques to re-engage those doing repetitive jobs:

- When engaged in a repetitious job, mentally and verbally treat it as if this were the first time performing the task. Verbally? Yes, audibly talking ourselves through the job helps our minds concentrate on the task afresh.
- Have the new employee train or talk another employee through the job.
- Stop and think: what could go wrong with this job, this time?
- Stop and think: is the procedure for this job correct at this time, under these conditions?
- Stop and think about how this job can be improved (including safety measures) – as long as proper protocol is observed.
  This kind of refocus technique is referred to as situational awareness: being able to focus on the job every time as if it were your first time performing the task. Verbalize how to do the job and engage the brain before stepping into the field of fire. It is all a part of the culture of correct: training our minds, actions, and employees how to do every job perfectly, every time. In this case, the training is not the responsibility of a supervisor or upper management, but rather the responsibility of each person doing the job, every time he/she does the job.

21.4 Safety Warnings – the Severity of the Outcome

Many of the safety warnings in our workplaces minimize the potential seriousness of an incident related to the warning signage or JSA document. This downplayed evaluation of such dangers gives employees a false sense of security and reduces their personal risk awareness and caution:

- Pinch points are more often crush points, or potentially fatal crush areas.
- Hot water is not the same temperature as bath water. Water that is 150 °F or above will scald, and can potentially kill those who come in contact with it.
- Gas warning signs do not adequately portray the danger of toxic fume inhalation, or the fatal effect when flammable gases are ignited.
- Fall protection signs do not convey the high probability of permanent disability, or death, from falls as short as 4 ft.
- High visibility clothing and spotters do not sufficiently reduce the risk of death by crushing.
- High voltage warning signs do not fully communicate the severity of the risk of electrocution.
- Confined space warnings do not portray the numerous deaths which have occurred as a result of this work area danger.
- The warning of a potential trench collapse fails to represent the number of deaths by suffocation which occur every year from this very real danger.

Many other situations can be added to this list of workplace dangers that are not described to the full magnitude of their potentially tragic consequences. Such an abbreviated description can cause employees to let their guard down and expose themselves to dangerous safety risks.

How can safety signage shortfalls be combated? Warning signs, safety videos, and computer programmed safety training do not adequately educate employees on the potential safety dangers indicated by warning signage. A far more effective technique is frontline training by individuals who have been, or are currently, exposed to the safety dangers. There is also value in learning from individuals who have experienced safety-related tragedies first hand. For many of the above risks, an impactful story would heighten awareness and decrease the risk tolerance of those who face such workplace dangers.

The Doc once worked with a mining company that constructed a list of Fatal Risk Protocols. Each year, every employee received extensive training in 11 underground issues that had time and again led to fatalities. The concept of this training reflects "Stop and Think." What could happen, and how severe are the consequences?

A robust safety culture must know what safety risks exist in the workplace, and how leadership can effectively communicate and knowledgeably train the potential severity of the outcomes. Consider the value of having employees continually note/list safety hazards that they encounter in the workplace and then using these as topics for continuous improvement teams to resolve.

21.5   Voluntary Actions and Being in Control of Them

We all do voluntary off the job activities that include risks. It is normal for our risk tolerance to increase for these personal non job-related, mostly enjoyable times in our lives. With increased risk comes increased probability of injury. Some time ago a National Safety Council (NSC) report commented that about 70% of all medical case injuries occur off the job along with about 90% of fatal injuries.

Nine out of 10 deaths and about 70% of the medically consulted injuries suffered by workers in 2010 occurred off the job. While nearly 14 times the number of deaths occur off the job compared to on the job (13.8 to 1), more than twice as many medically consulted injuries occur off the job (2.7 to 1).

Source: National Safety Council estimates.

This factor brings to mind a quote from Albert Einstein: "There are only two infinites; the universe and human stupidity. I am not all that sure of the former." Think about typical off the job activities, like riding a motorcycle. How many of you would consider your company trading in trucks for high powered, no seat belt, two wheel cycles? Think about your own off the job recreation and the risks involved with this type of experience. Are you exposed to much higher risk with much lower personal protection? Take some time to gather your thoughts about your off the job activities and risks which are unthinkable on the job.

What should be done with this kind of personal risk? Dave Fennell, one of the more forward thinking safety professionals whom the Doc has had the pleasure of working with, initiated a "Stop and Think Card" to help with personal risk assessment for both on and off the job activities. The laminated wallet-sized card prompts people to ask:

- What is the scope of what I am about to do?
- How could this scope change to become more dangerous?
- What could go wrong?
- How bad could it be?
- Do I clearly understand my tasks?
- Am I physically and mentally prepared?
- How can I reduce the personal risks to myself and others?
- What would the effect be to my family?

Before you head to the fields and the woods for your weekly farm chores, or other off the job duties, pull out this card; stop and think as you verbally ask yourself these questions and actively work to reduce your personal risks. The approach does not pertain to just higher risk-related activities, it needs to become a habit that makes sense in many other activities. The personal risk reduction concept here is akin to a personal tool box talk each and every time you are about to begin a risk-related task on or off the job. You will truly have a noticeably higher personal peace of mind and lesser potential risk of injury by doing so. Why not consider a similar personal commitment on your part for you, your family, and your at work organization?

21.6 Personal Experience with an Outcome

Essentially each organization has dangerous circumstances that can maim or kill an employee. Many, many organizations have experienced events that have maimed or killed a fellow employee or contractor. One of the noticeable mechanisms to reduce personal risk tolerance is the personal memory of a tragedy which influences us to be more careful, to reduce our own personal tolerance and acceptance of the associated risks.

While working in the underground hard rock mines of South Africa, the Doc was introduced to an organization's 11 Fatal Risk Protocols (FRPs): conditions and operations that could most definitely and quickly kill anyone in their daily work environment. The safety director, Mohlaba, then described his father's death as a result of one of the more common FRPs. As a result of Mohlaba's personal story, when the Doc went underground he definitely paid close attention. Thinking back, it was evident that all 11 Fatal Risk Protocols had similar tragic stories behind them, and that all the underground staff would have benefitted from similar corporate knowledge storytelling.

Somewhat farther back in the Doc's life, he was operations director at an explosives manufacturing facility. On separate occasions, individual employees related their personal experiences around fatalities which had occurred with the processes this company ran on a regular basis. Their intent was not to shock the Doc, rather it was to point out the dangers of what looked to them to be a slippery slope that people were getting too close to, and that must be avoided.

The Doc related that he does not ride motorcycles because years back, while doing so, a car took a sudden left turn in front of him and he went through the car's windshield. Yes, riding motorcycles was fun, but the risk was now too high for the Doc based on his personal experience and the near death of his friend, who was riding behind him on the motorcycle.

Each year in the United States there are approximately 5000 industrial work-related deaths, about 50 000 vehicle-related deaths, and about 60 000 off the job, non vehicle related deaths. The number of medical cases for similar events is far larger. Each of these has personal stories which can be effectively related in the workplace in an effort to reduce the risk tolerance of ourselves and others. No matter where you live or work, developing a practice of regularly retelling safety event history can help you and your organization reduce injuries both on and off the job. People like to reminisce and tell stories so why not give this approach a try?

21.7   Cost of Noncompliance

A high cost of not complying with the rules definitely affects our personal risk tolerance. An example that comes to mind deals with airline travel. A recent flight delay announcement was for "a minor maintenance item that has been taken care of and must also have the paperwork completed before the flight can begin." The cost of noncompliance with issues and procedures in the commercial airline industry is too great to take chances. Even small risks in this industry are not tolerated.

The other end of the travel spectrum risk tolerance experience came when the Doc drove the traffic clogged expressways of the Boston, Massachusetts area. When the Doc and his family lived near Boston, they vividly remember their insurance agent handing them a copy of the book *Wild in the Streets, a guide to survival while driving in Boston.* There were adequate rules of the road, they were just neither followed, nor enforced to any significant degree while he and his family lived there. As a result, the insurance agent was careful to point out the risks the family needed to be aware of so that they could act accordingly.

In industry it is common for high cost of noncompliance issues to be associated with policies and procedures which are known as Cardinal Rules. The potential serious injury consequences of breaking a Cardinal Rule are so high that there is no tolerance for employee noncompliance at any level of the organization. Break one of these Cardinal Rules and an employee will be let go to find another job someplace else, with some other company. Examples of these Cardinal Rules include items such as:

- Lock Out Tag Out (LOTO) associated with large multi stand metal rolling mill operations and maintenance.
- Confined space work associated with the chemical process industries.
- Fall protection utilization on Arial Work Platforms in the construction industry.
- Substance abuse in oil field drilling and other industry's operations.
- Texting while driving.

There need to be both on the job and off the job Cardinal Rules which force us and our associates to significantly reduce our own personal risk tolerance because the potential consequences of noncompliance are just too great. Think about the Cardinal Safety Rules that must be followed in your industry, and in your off the job life.

21.8   Overconfidence in the Equipment

Over the recent years there have been significant advances in technology which have reduced injury risk. Some examples include:

- Bucket trucks which give better, more stable access at heights.
- Automatic Braking Systems (ABS) which help vehicles to stop more quickly and safely.
- Ground Fault Interrupting (GFI) electronics which reduce electrocution risk.

There are many more such advances that have reduced injury risks in what people do both on and off the job. However, a human phenomenon called *risk homeostasis* commonly comes into play; as technology reduces risk, we humans have a tendency to increase our personal risk tolerance, often negating the benefit of the technology assistance. A couple of examples of risk homeostasis include:

- An increase in aggressive driving as a result of more confidence in an ABS (Anti-Lock Braking System) being able to safely stop the vehicle.
- Little or no change in parachute sport fatalities, even though the equipment failures have been reduced dramatically. Now the predominant fatality cause has shifted from no chute deployment (equipment failure) to delayed chute deployment (personal decision to delay chute deployment due to overconfidence in the equipment working properly).

Think about the technology improvements and increased confidence which have resulted from computer controls, warning systems, hardware improvements, and the like in industry. How has technology advancement lulled our workforce into more confidence? How has this increased our overconfidence and thus increased our organization's employee tolerance for personal risk? Each of us can make a list of areas where we are more at risk as a result of increased technology. This is a good safety exercise for the people in all organizations.

What is the solution to increased personal risk tolerance due to overconfidence in the equipment? There can also be increased risk associated with overconfidence in automated control systems such as an ABS coupled with a lack of understanding of how they function. As an example, understanding of ABS and other systems is very beneficial in case the system fails or its "logic" is not fail-safe.

One of the more effective approaches goes right back to the stop and think card. Before each job, stop and think and discuss:

- What could go wrong?
- How serious could it be?
- What can my crew members and I take personal responsibility for to reduce our personal risks for which we have become more tolerant?

The old hands and the new employees need to stop, think, discuss, and thus decide to reduce their personal risks; as if their personal actions, not technology, were responsible for their own personal safety.

21.9   Overconfidence in Protection and Rescue

A corollary to overconfidence in equipment is having too much confidence in personal protective equipment (PPE). Technology advances, in some cases, have led us to be more confident in PPE preventing injuries, thus increasing our personal risk tolerances to the dangers around us. Some examples include:

- Fire-resistant clothing (FRC) that does not burn when exposed to flame: When wearing fire-resistant clothing, individuals may take risks that they would have not taken otherwise. While fire-resistant clothing (e.g. Nomex) does not burn, it does get very hot and can still burn the employee wearing the protective gear.
- Gas detectors designed to warn of the presence of dangerous gases: Gas detectors are designed to detect specific gases. For example, H2S monitors do not alarm in the presence of flammable gases which do not contain H2S. Other dangerous gases go undetected.
- Impact-resistant gloves designed to protect the hands: While impact-resistant gloves protect individuals from minor forces of impact, they do not protect against high impact forces that can crush or amputate. (Remember that "impact resistant" does not equal "impact proof.")

To prevent decreased personal risk tolerance, an individual should work as if unprotected. Working as if exposed to various levels of safety risk can help individuals take accountability for their safety rather than solely relying on the technology of PPE. By identifying and understanding the limitations of PPE, one can:

- Identify the risk
- Mitigate the dangers
- Improve functionalities of safety systems
- Use PPE as a final line of defense.

Another common worksite overconfidence is relying upon rescue and medical technology. The real deterrent to this overconfidence, which leads to higher personal risk, is to understand the limitations of protection and rescue measures. As an example, air and ambulance evacuation for remote or even local sites can take too long to prevent the serious results of an injury.

Without a doubt, the improved technologies which exist are meant to be the LAST line of defense. We must:

1. Try to eliminate the risk.
2. Mitigate the dangers.
3. Improve our safety systems' functionalities'.
4. Leave PPE and medical treatment as a last line of defense, not something to be relied upon.

Pre job briefing that presents all employees as becoming older and therefore potentially less capable and unprotected to the dangers we can encounter is a good concept. Presenting this reality before we step into work and off the job-related risk environments will bring some realities to the forefront of one's mind, especially if they rely too much on rescue and medical technology.

21.10   Potential Profit and Gain from Action

There is a direct correlation between injuries and associated economic pressures. As the economy improves, there is a clear benefit to getting one last load delivered. As the price of raw materials increases, there is a natural tendency to take additional risks necessary to get more volume produced. As the end of the calendar quarter approaches, there is a push to get out every last item promised or possible. As downtime increases, maintenance safe guards are compromised to get the line back up ASAP (As Soon As Possible).

There are personal economic benefits which drive us and our employees to focus on delivering more and more, often taking shortcuts. Faster then becomes more important than safer. The resultant increased personal risk tolerance leads inevitably to increased injury rates. Promise dates and delivery deadlines can noticeably compromise safe operation necessities. We need to consider the effects of taking/not taking shortcuts which may reduce the amount that can be produced.

Is your organization mature enough to remove rewards for risk taking, such as per day and per unit incentives, which, in turn, compromise safety for you and your employees? How can the organization help eliminate these behind the scenes barriers to "doing it the right way," rather than living a higher potential injury culture of "get 'er done," or "just win baby," or delivering every possible last dollar to the bottom line? Think about process from an analytical viewpoint; how does the potential marginal return on investment for taking these risks compare to the personal trauma and costs associated with potential injury or damage in doing so?

Many of these kinds of issues have been a part of individual organizations and industry for generations. Leadership often does not even realize they exist until they collectively stop and think about the profit message they are sending and living while neglecting and safety-related factors. Such tacit messages lead to higher personal risk tolerance and injury rates, which in reality, are directly opposed to higher return on investment rates. Where does your organization need to stop and think, and then act, about legacy systems which send the wrong personal risk assessment and risk taking messages?

21.11 Role Models Accepting Risk

One of the Doc's ongoing frustrations with many organizations is their inability to effectively address and deal with notable role model personnel who continue to take significant risks. Sometimes it is:

- Inexperienced work groups which are just trying to get the job done without adequate operations or safety training and knowledge.
- Tough hourly organization personnel who resist any kind of change as a supposed threat to their power. An extreme case that comes to mind was the AK Steel case that Jim Stanley and Dr. Dan Petersen were so prominently involved with a number of years ago (*available with an internet search*).
- Management leadership for which the focus on output trumps all other considerations.
- The older hands who have always done it in a way that is no longer acceptable after implementing a safety culture that is more focused on preventing the possibility of injuries.
- A pet peeve that is a poor safety leadership example which just keeps occurring.

Sometimes these key poor safety leadership personnel are punished, or rewarded, or what seems most often the case, ignored, even though their performance with respect to safety affects the decisions and actions of all the others within their company. Experience shows that sometimes there are no injuries which result from this kind of poor leadership, whereas tragic injuries and fatalities can also be the result.

The Doc mentioned that in his mind, the real issue with role models who continue to take risks seems to nearly always deal with inconsistent and/or ineffective safety leadership which does not appropriately address the effects of poor role models.

- Do we really know who the true safety role models are and what they are doing?
- Do we immediately address the unsafe issue and act to correct them?
- Do we counsel the poor safety role model personnel in an appropriate adult manner?
- Do we place good safety role models in responsible charge and then reinforce their decisions to make the low safety risk approach the correct action that is identified and publicized to assure its visibility?
- Are we visible in the workplace reinforcing the safe and productive culture?
- Do we reinforce the need for the work groups to stop whenever there is a safety concern, discuss the situation, and then choose the safe, low risk alternatives? This is not a search for an excuse not to do a job, rather it is a search for ways to do the tasks safely

A poor safety role model culture requires strong safety leadership commitment and accountabilities from personnel who continually engage in creating and demanding a correct safety culture.

Considering this Personal Risk Assessment material, there is an important *fourth pillar* of safety performance:

1. Regulations that set the basics in safely reacting to the conditions we face in the workplace.
2. Accountabilities for personnel at all levels and throughout the organization for engaging in what they can do to prevent injuries.
3. Error proofing the safety processes by engaging our people in a continuous improvement team safety culture.
4. Personal Risk Assessment that leads to a safety culture for all personnel who, as a result, just does not take risks.

The work Dave Fennell of Imperial Oil ExxonMobil Canada has done on this topic has definitely forced the Doc to consider an entirely different outlook on personal and corporate safety reality. Personal risk assessment starts with three common threads to analyzing personal risk, which all of us must consider:

1. Recognizing the hazard that is before us.
2. Understanding why it is a risk.
3. Evaluating our personal risk tolerance when we face the challenges of what we are about to do (Figure 21.1).

This process, in turn, leads to identifying 10 common personal risks and what individuals and their work group personnel should do when faced with these 10 deadly safety risk issues:

1. Overestimating our capabilities
2. Familiarity with the task
3. Seriousness of the potential outcomes
4. Voluntary actions
5. Personal experience
6. Cost of noncompliance
7. Confidence in the equipment
8. Confidence in protection
9. Potential profit and gain
10. Role models accepting risk.

Included with each of these personal risk topics are suggestions on how to reduce the risks that individuals and their work groups experience on a daily basis. As an example, consider the "stop and think" process/card which, when used on a regular basis, causes our employees to consider the risks, how serious

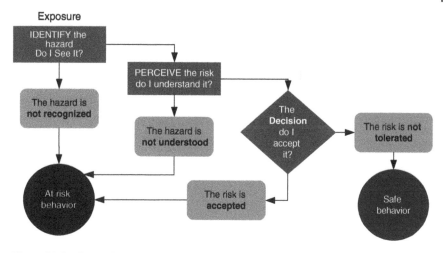

Figure 21.1 Perception and tolerance model. *Source*: Dave Fennel.

they could be, what they can personally do about them and more. This is a practical tool set the Doc now carries with him wherever he goes in dealing with safety issues. Part of the real power of this simple tool is delivered when leadership and personnel, from across the organization, continue to reinforce its use by all people just as a part of being alive, like eating or breathing. The desire is to create a personal risk assessment culture which lives as "This is just the way we do things around here on and off the job."

If you deliver this kind of culture, you and your people will need to:

- Study the above mentioned personal risks and their tool sets.
- Customize them to your organization's reality.
- Train and ***continually reinforce*** personal risk assessment with appropriate action accountabilities as an important part of your organization's safety culture practices from top management through hourly employees.

> *It will be a long session that requires* possibly two days of *training when Aaron's company presents it to the employees. Aaron and his fellow newly empowered safety leaders are reeling from all the material and the implications that directly affect their company's safety culture and the livelihood of all their employees. Aaron detects* similar shocked looks coming from *the others who have been a part of the Doc's intense sessions on safety culture excellence. A hand is raised, and the question asked: "How are we going to accomplish all this? We are overwhelmed!"*

> *The Doc smiles and lays on some more of his Papa's wit:* **"Don't pray for rain, if you aren't willing to play in the mud."** "*After the break, let's do a brief wrap up to answer that question in more detail."*

22

The Trap of Complacency

Aaron is excited as he and his Safety Steering team (SST) return after the break. "We are finally nearing the finish and the whole team sure appears to be engaged and supportive! The herd is assembled. Let's head 'em up, move 'em out and get our Zero at Risk actions show on the road!!" The Doc does an overview and briefly walks them through each of the models (shown throughout the Chapters of this book). He then explains the process of integrating all these models in a step by step manner which has the teams solving the organization's problems one at a time. The real purpose of this safety turnaround material is the process of how an organization successfully implements each of the steps with their own people. These are deep dives into complex people and systems problems that have been around their organization for years.

One of the objectives of the Doc's team is to train (better yet educate) the people of an organization how to fix its own problems using their own people. This result is accomplished by following the process; by learning how to fish, so to speak. However, leadership can't let the process stop before they have learned how to apply the models and have the ability to do so on their own, without requiring consulting support.

The Doc then explained the dangers of "good enough complacency" with respect to downstream indicators like RIF. The backsliding that comes from allowing this good enough complacency poison to live in an organization can absolutely nullify their efforts to achieve a culture of excellence. Good is the enemy of great just like putting lipstick on a pig does not change the pig.

A long while back, the Doc and his Papa had a few years of bonding as they bought an old car body and began a multi-year project to build the Doc's first automobile. First they did body work on the dents and holes. As they finished paint priming, the (much younger) Doc stepped back with a big smile, enjoying how good it all looked. And then one of his Papa's wisdom moments came into play as

Delivering Safety Excellence: Engagement Culture at Every Level, First Edition. Michael M. Williamsen.
© 2021 John Wiley & Sons, Inc. Published 2021 by John Wiley & Sons, Inc.

Papa said: "Son, it takes more than lipstick to make a pig beautiful." And for the next few years, they toured various wrecking yards to purchase old parts for the car, refurbish them, and finally get the ol' pig up and running. In the end, it sure was a good looking pig that went way beyond their original lipstick.

When the Doc's son, Erik, reached high school age, they spied a somewhat newer derelict car, bought it, and did their own bonding adventure over the next few years. Yes, the Doc did have opportunity to pass on his Papa's wisdom about pigs and lipstick. They had a number of good laughs and heartaches over his son's slowly beautifying pig. Once Erik got married, his wife thought they needed an even better looking and more reliable vehicle than Erik's "62 Willys wagon. And so Erik's beautiful project pig now rests under a tree on the Doc's small farm. From time to time they circle the ol' Willys pig and reminisce about these challenging, personal growth times they experienced together.

Truly, it took a lot more than lipstick to get a much neglected derelict vehicle to become serviceable and safe. A significant makeover was needed, which required time, money, ongoing effort, and a commitment to see the work through to completion. Many safety pros have stepped into a new job situation and been faced with a safety pig that was truly ugly to behold. Likely, they all made a decision where to start and, on completion of the first round of efforts, they then stepped back and marveled at the noticeable improvements in their safety pig.

To get to a finished safety culture that just does not have incidents or injuries, you will have to go way beyond the first lipstick effort which brings a smile to your face, but barely begins to positively affect the people and company culture for which you have responsibility. The true safety pros are committed for the long run of what is required to turn a pig into a beauty queen. Do not call it quits and move on to the next safety pig without completing the multi-year project which develops the zero incident/at risk activities safety culture your people need for them and their company to work injury free.

Sometime later, Aaron attended a retirement party for one of his frontline friends, Lynn, who was injured in a 40 ft fall at his former company. The employee talked about the medical technology advances in the last few decades which are nothing short of astounding. Many surgeries and treatments which used to require extensive hospital stays are now day surgeries, performed in a doctor's office, and the patient returns home relatively quickly. To mention just a few of the amazing medical applications: plastic surgery removes scars, titanium replaces bones, tape replaces sutures, stents take the place of vein removal and relocation, etc. This former lineman was employed at an electrical utility which seriously worked on improving their injury rates. Lynn is one of those impressive personalities who is not satisfied with status quo. He was personally involved and engaged with the hourly and salaried employees in the quest for zero injuries/at risk activities. Lynn talked about himself and another senior lineman who had been at the top of their trade. Each was well respected, each had high confidence in their skill and experience, each had fallen 40 or more feet from a tower, each had amazingly survived, each had received excellent medical treatment and recuperated, each had gone back to climbing towers. And no matter the medical technology: the broken back, the forever asleep and numb hand, the age-related arthritis as a result of broken bones, etc. still persisted in his daily routine. All these painful physical reminders of their past on the job injury remained a part of their current and future 24/7 lives plus their views on employing/living essential safety procedures.

In the conversation, Lynn provided a serious insight which convinced Aaron that: no matter what medical technology is applied to you, the OEM (Original Equipment Manufacturer) body parts, their function and their feel, are all significantly better than the "aftermarket manufacturer" medical technology-installed parts Lynn now has to live with. Lynn's sharing of his personal story convinced Aaron that he must help his people realize their personal need for taking no short-cuts of convenience, no matter what their individual skills and experience are, no matter what stress there is to get the job done right now!

Aaron thanked Lynn for his insights and promised to bring him back from time to time to share his safety story and life messages during new employee orientations.

As Aaron drove home, he thought about the safety department and the organization's safety culture five years in the future. Being committed to executing the purpose, the process and the models will make a huge impact on the costs incurred by a company and the morale boost from everyone going home safely every day This same paradigm shift also impacts the reality of Aaron's professional safety future, and the livelihoods of the 1000+ employees who work in the organization where Aaron leads a zero at risk safety culture. In fact, the new CEO and the new general manager, who have both been actively supporting Aaron's safety culture excellence initiative, are having a ripple effect on improving the safety numbers for downstream employees. Hourly employees are increasingly becoming more actively involved in fixing safety-related problems they used to just live with. In addition, they are applying peer influence to others who used to take dangerous shortcuts. In just 18 months, the RIF has dropped to less than five while the DART rate is down by about 60%, and costs are way, way down. Off the job injuries are also noticeably down.

Once home, Aaron smiles as he realizes his many work friends' personal pain is way down as injury rates continue to decrease. Aaron reflects on another truth of one of his own personal clichés which helped this work take hold: "People do not care how much you know until they know how much you care." The care actively exhibited by people at all levels of the organization has rippled throughout, and led to all employees being a part of this important safety culture turnaround.

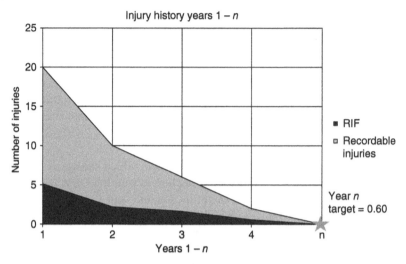

Figure 22.1 Lagging indicators.

Year	RIF	Recordable injuries
1	5.12	20
2	2.2	10
3	1.64	6
4	0.58	2

However, also in his mental background, Aaron knows the Doc has given the organization a valuable caveat: They must not become complacent with their relentless pursuit of zero. The RIF and other trailing indicators are all better, but nowhere near BOB (the Best of the Best) the organization wants to achieve.

Hopefully Aaron and his teams' ongoing efforts will lead to as successful result as that of an even larger global heavy industry company which uses the Continuous Excellence Performance (CEP) process shown in Figure 22.1.

Epilogue

Some Final Thoughts from the Doc

The Jelly Layer: Front line safety

One of the safety pros I (the Doc) had the pleasure to work with describes the front-line work environment as the "Jelly Layer." Sure, companies have absolutes when it comes to values and principles such as Ethics, Safety, Quality, Cost, Schedule . . . . When you are facing many multiple inter-relations at the front line, solutions are not always simple. You are in the Jelly Layer, and all the regulations, policies, procedures, principles, etc. seem often to deliver some level of conflict, and compromises begin to creep in.

Many of us have been a part of the Jelly Layer more than once in our careers. Add to the uncertainties and conflicting priorities, plus the dynamic of a demanding and/or authoritarian leadership consisting of multiple people who are not always in agreement and our Jelly Layer gets even more complex and dangerous to our employees, ourselves, and our careers. We all get to juggle more than one ball at a time in our life spans. On occasion, we all have more balls in the air than we can successfully handle, and inevitable ball drops begin to occur. If/when this occurs it is time to make a personal risk assessment which guides us to an at the moment unpleasant decision about which failure (compromise) to live with.

A strong team, with multiple talents to assist us, definitely leads to fewer drops. To successfully play this Jelly Layer endgame you must have carefully considered your personal values and be prepared to actively live them at a moment's notice. There is no opportunity nor time for navel gazing in the Jelly Layer. However, right now, there is time for some personal reflection and commitment to your own personal values.

The story you read in this book began when I became deeply involved in helping organizations achieve safety culture excellence back in the 1980s. At that point in my career, I was in charge of manufacturing engineering for a Fortune 20 company. One of the facilities experienced a fatality which, in turn, led to

Delivering Safety Excellence: Engagement Culture at Every Level, First Edition. Michael M. Williamsen.
© 2021 John Wiley & Sons, Inc. Published 2021 by John Wiley & Sons, Inc.

a serious corporate-wide safety initiative. Within a week of the tragedy, I was assigned responsibility for the safety of about 10 000 manufacturing employees who worked at 40 plants scattered across America. The organization ran an average recordable injury rate in the 20's and had no real safety focus. Production was king, as measured by upstream indicators of cost, quality, and customer service.

After some research, the company hired the famous safety consultant, Dr. Dan Petersen. For three years Dan and our team lived on-the-road developing a safety culture excellence model which quickly delivered a corporate-wide average recordable rate (RIF) of about 1.2. We were all very happy with a safety accountability culture which worked well across the entire company scattered across the nation.

Following this success, everyone went back to their "normal day job." After all, we were convinced we had accomplished the mission for safety. The turnaround leadership and I then left the company for other, "greener pastures." Some years later, after presenting the success story at a safety conference, the then current safety manager for that organization stopped me to discuss his current safety situation. It turns out that five years after we all celebrated how good we were, the whole organization was back at about a 20 recordable incident rate. My new safety culture had stuck for a short while after the resources were reallocated to the ever dominant production culture, and then collapsed.

My lesson from this round trip, and the many difficulties in sustaining safety culture excellence is, you must **never** declare victory. The relentless pursuit of zero, must be relentless. If we back off, the second law of thermodynamics effect (all things degrade over time) seems to take over and the entire system degrades. This "system degradation" is exactly what our production and quality culture counterparts have learned. We all must consistently keep alive the push for excellence. To quote Yoda, "There is no try, there is only do." We must stay engaged.

"Life is what you make of it – always has been, always will be. There will always be challenges. It is what we do with them that counts. I have written my life in small sketches." The above quote from famous New England painter Grandma Moses (1860–1961) has some real truth and a challenge for the safety professionals of our era. As I look back over a "career" that includes some 20 organizations, there are many sketches, certainly not all are success stories. When I talk to others who have stayed with one organization for their "real career" they too have the all too familiar ups and downs.

If we add to the work environment family concerns, off the job experiences, volunteer work, and the like there seems to be a nearly infinite number of sketches which help weave together the real "who" we end up being. Few I talk to are settled into the grand success plan they dreamed they would have upon graduating from whatever school/grade they were a part of years ago; are you?

We still have some sketches left to paint. As a safety pro, and even more so as a person, how can we make these final sketches clear, long lasting, satisfying, and of value to the others we interact with on and off the job, now and down the road? Ongoing employment or retirement, even if it leads to a time user position being employed as a big box store greeter or a volunteer for some organization we care about, can have meaningful value to you and those with whom you interact. What is needed is for us to focus on delivering a meaningful personal sketch in what we are doing in our current stage of life.

May you choose to join us in the endeavor of: developing a culture of excellence – the long, enervating journey to a zero incident/zero at risk safety activities safety culture. It is a ZIP line adventure that will provide you with ongoing enjoyment and benefit.

The Doc

A

The History of the Continuous Excellence Performance (CEP)/Zero Incident Performance (ZIP) Process

The Continuous Excellence Performance (CEP) process has its foundation in the work of Dr. W. Edwards Deming. Dr. Deming is widely credited as the architect of the quality revolution in Japan following WWII, and later in the United States and other parts of the world. Dr. Deming developed a process improvement methodology that has found common use in business and industry today. In the 1980s, Dr. Dan Petersen, considered to be Deming's counterpart in the field of safety, began to apply the quality methodology to safety. In response to requests from the railroad industry, Dr. Petersen and Dr. Charles Bailey developed and statistically validated a safety perception survey to assess the railroad industry's safety culture for the purpose of improving it (Appendix B). Meanwhile, Mike Williamsen was working for a Fortune 500 company which had become quite concerned about its safety culture following a severe incident. At the time the company had amassed a recordable incident rate around 20 and assigned Mike Williamsen to study the problem and work to improve the safety culture. His study led him to Dr. Petersen, whom he hired, and the two of them became "road travel buddies" as they worked throughout the United States to figure out solutions for the company's safety problems. Mike Williamsen, having come to the journey with a strong process improvement background, began to incorporate Dr. Deming's methodologies with Dr. Petersen's experiences in formulating an approach to the company's safety incident/culture issues.

During the same time period, a husband and wife business team had formulated a safety products company called CoreMedia Training Solutions, in Portland, Oregon. They were busy developing safety education and training improvement products and offering them to the American business market. The wife, Katy, had worked for several years in the burn unit of a hospital, witnessing the horrible consequences that resulted from injuries in the workplace. Consequently, she developed a sincere passion to improve workplace safety. The first products the CoreMedia developed focused on safety compliance. Eventually, however, they realized that the main problem with safety in US industries was not in

Delivering Safety Excellence: Engagement Culture at Every Level, First Edition. Michael M. Williamsen.
© 2021 John Wiley & Sons, Inc. Published 2021 by John Wiley & Sons, Inc.

compliance, but in the area of safety culture. As their thinking progressed, they began to understand that most people take risks not because they want to get hurt, but because they think these risky actions are what their boss or their organization wants them to do to improve costs, or production, or some other issue. Since Katy had become convinced that culture was the main problem in safety, she set out to learn all they could about it. Her learning journey led to the work of Dr. Dan Petersen. After studying Dr. Petersen's work, she entered into a business arrangement with him to productize his safety culture methodologies. Their early joint venture products included *The Safety Management Series*, *Safety Accountability*, *The Challenge of Change*, and the book, *Authentic Involvement*, all of which contain legacy materials and concepts that still apply to today's safety culture realities.

In the late 1990s, they decided to transfer CoreMedia to their son, Tim, and pursue other passions. Tim, an entrepreneur sort, soon expanded the company's focus beyond just safety products to also include safety consulting. In 2001, Tim hired Dr. Mike Williamsen (BS Chemical Engineering, MBA, PhD in business and the author of this book) to help him build the consulting side of the business. Dr. Williamsen began to pull together the Safety Perception Survey, the Dr. Petersen products, as well as various and diverse safety improvement technologies. The team's significant efforts allowed CoreMedia to begin selling Dr. Petersen's safety perception survey to companies as the introduction to consulting services and safety products. In 2006, Tim partnered with a sales expert, Shel Reed, to improve their ability to sell and market their services into dangerous workplaces. Shortly thereafter, he hired two more consultants, Todd Efird and Todd Britten, and the expanded team began working with companies to improve their safety cultures. To figure out how to best sell this service, Reed posed this question to the group: "What do we call this thing we're doing?" Reed noticed them referencing continuous improvement, safety culture, and leadership engagement, and he wondered how all of it fit together. The diverse skill sets of the team enabled them to brainstorm many options and approaches to organizing the material. Todd Britten suggested the name be changed from Continuous Excellence Performance (CEP) to "Zero Incident Performance (ZIP)." The first customer to purchase the entire ZIP Process was a large regional electric utility in Canada. This same "ZIP Process" is now in use worldwide in many organizations and industries. However, since "ZIP" has become a trademarked item this book uses the original term Continuous Excellence Performance (CEP) as the title/header for the explanation of the safety culture improvement process basics.

The roots of Continuous Excellence Performance/Zero Incident Performance are in the process improvement work of Dr. W. Edwards Deming and the significant pioneering safety culture work of a Deming era peer, Dr. Dan Petersen. Tim and the CoreMedia team used these foundational materials to develop a methodology for safety culture improvement that has proven valuable to many companies in many industries worldwide. Since that formulation in 2006, the "ZIP / CEP Process" has remained intact, and has also been improved by numerous continuous improvement teams. These dynamic materials served as the foundation of the consulting services then provided by CoreMedia.

In 2011, Tim sold CoreMedia to a global industrial manufacturing concern and he began managing the business as an added service within the global company's culture. The team grew from its small original group to more than 50 employees serving scores of companies and thousands of employees worldwide. As of this writing (2020), the group continues to adjust and improve as the team revisits the "ZIP/CEP" Process methodology. These adjustments to the process have also included new safety culture processes, which have been added as a result of acquisitions in safety technology and fatigue management. As these and other needed improvements occur, the core of the "ZIP/CEP" Process methodology remains foundational to their work.

It is important to note that while the information contained in this book is valuable, it is only a guideline of how to engage in the "ZIP/CEP" Process. The most important factor for organizations needing safety improvements is how they can deliver "ZIP/CEP." Just as Toyota leadership said to General Motors as they toured Toyota facilities to learn about quality management, "You can take all the pictures you want, because we know the secret to our success is not in what we do, but in how we do it." This same concept applies to organizations needing to improve their safety cultures. The *what* (models presented in this book) is important, but it is the *how (the story line dialogue in the book)* which creates the true value to the organizations in need. How you, the reader and your organization, deliver the process is essential to improve your safety culture! May you be successful in transforming your safety culture to "ZIP/CEP."

B

The Railroad Study by Petersen and Bailey

{My son cleaned this up. I put it in the appendices as documentation of how the whole process started. I think it is a good history for this book and also a good testimony to two great safety pioneers who truly started the cultural approach to safety. I am not sure this material is discoverable much of any place else. I am not sure of the need to provide reference since it all came from a photo copy of the booklet CoreMedia had that was written by Bailey.}

Using Behavioral Techniques to Improve Safety Program Effectiveness

by Charles W. Bailey

Delivering Safety Excellence: Engagement Culture at Every Level, First Edition. Michael M. Williamsen.
© 2021 John Wiley & Sons, Inc. Published 2021 by John Wiley & Sons, Inc.

Based on a Study Conducted for the Safety Section of the Association of American Railroads and the Federal Railroad Administration by
  Nicholas J. Andrews
  Charles W. Bailey Robert F. Bain
  Jack C. Buckingham Thomas M. Hatchard
  Frank M. Kaylor
  Rolland G. Kuhlmann
  Dan Petersen
  William D. Underwood
This study was performed pursuant to contract number DTFR 53-85-C-00039 awarded by the Federal Railroad Administration, U.S. Department of Transportation to the Association of American Railroads. Points of view and interpretations of study findings stated in this document are those of the author and do not necessarily represent the official position or policies of the abovementioned organizations.

B.1   MR Study of Safety Program Effectiveness

B.1.1   Phase I – 1979–1983

SAFETY PROGRAM ACTIVITIES SURVEY – Determined safety program organization and activities on 18 major railroads, scores developed by Aberdeen Group for 12 factors thought to be indicative of safety, program success and matched with performance data for each railroad.

  PILOT SURVEY OF SAFETY PROGRAM EFFECTIVENESS – RAILROADS I and II – Questions concerning all aspects of safety program effectiveness, from both behavioral and procedural/engineered standpoints, asked of five organizational levels from hourly rated employees to executive management on two railroads AAR Study Group analyzed response, discarded questions with no statistical significance, and used remaining questions to develop questionnaire used in study of two additional railroads in phase II.

  SUPERVISOR USE OF POSTIIVE REINFORCEMENT – RAILROADS I and II

  BASELINE OBSERVATIONS OF EMPLOYEE BEHAVIOR – Trained observers using standard lists of unsafe behaviors. Observations made on both study and control divisions.

  SUPERVISOR TRAINING – On study divisions, taught positive reinforcement techniques and put them into practice.

  THREE-MONTH OBSERVATIONS – Second round of observations made on both study and control divisions to determine effect of training on worker behavior.

SUPERVISOR TRAINING – On study divisions, taught employee assessment techniques and put them into practice.

SIX-MONTH OBSERVATIONS – Final round of observations on both study and control divisions to determine effects of training on worker behavior.

B.1.2 PHASE II – 1985–1988

SURVEY OF SAFETY PROGRAM EFFECTIVENESS – RAILROADS III and IV – Survey of five organization levels on two additional railroads using revised questionnaire developed in Phase I.

ANALYSIS OF SURVEY RESULTS – Development of analysis by AAR Study Group. Presentation to management to provide basis for safety program improvement.

SUPERVISOR USE OF POSITIVE REINFORCEMENT – RAILROADS III and IV

BASELINE OBSERVATIONS OF EMPLOYEE BEHAVIOR – Trained observer using standard lists of unsafe behaviors. Observations made on both, study and control divisions.

SUPERVISOR TRAINING – On study divisions, taught positive reinforcement techniques and put them into practice.

THREE-MONTH OBSERVATIONS – Second round of observations made on both study and control divisions to determine effect of training on worker behavior.

SUPERVISOR TRAINING – On study divisions, taught employee assessment techniques and put them into practice.

SIX-MONTH OBSERVATIONS – Final round of observations on both study and control divisions to determine effects of training on worker behavior.

B.1.2.1 Study Overview

This is a report on a study of railroad safety program effectiveness conducted in two phases on four railroads over a 10-year period. Each phase of the study looked at railroad safety programs with three evaluative tools:

1. *An inventory of the entire organization's opinions about the railroad's safety program and practices.*
2. *An experiment attempt to modify safety-related behavior by encouraging supervisors to practice positive reinforcement of safe behavior and other behavioral techniques.*
3. *An evaluation of change in safety-related behavior through a system of behavioral observations.*

In the first phase, a pilot study was conducted on two railroads from 1979 to 1983. In the second phase, experience gained through the pilot study was applied on two additional railroads from 1985 to 1988.

This report details results obtained through the study which led to the following conclusions:

1. The 12 most commonly used criteria in standard safety program audits are poor measures of program effectiveness.
2. Application of the 20 category survey technique developed by this study provides a reliable measure of safety program effectiveness.
   a. The AAR survey effectively identifies the strengths and weaknesses of the 20 management system areas affecting safety performance.
   b. The AAR survey effectively identifies differences in perceptions of the effectiveness of safety program elements between hourly rated employees and various levels of management.
   c. The AAR survey will effectively measure improvement and deterioration of safety program effectiveness when administered periodically.
3. Use of positive reinforcement techniques will enhance safety program effectiveness.
4. Behavior observation, properly administered, is an effective tool for measuring the effects of applying specific management techniques in the railroad setting.

B.1.2.2 Participants in Study

The study was conducted by a special study group appointed by the Safety Section Steering Committee of the Association of American Railroads. It was partially funded through two contracts with the Federal Railroad Administration.

The initial or "pilot study" (1979–1983) was conducted under a contract with the Office of Rail Safety Research with support from safety experts at the Ballistics Research Laboratory at Aberdeen Proving Grounds and The National Space Technology Laboratory, and from statisticians from Computer Sciences Corporation. The training phases were conducted in cooperation with faculty of the Masters of Industrial Safety Program at the University of Minnesota – Duluth.

The final study (1985–1988) was conducted by the Association of American Railroads Study Group under a contract with the Federal Railroad Administration. The Masters of Industrial Safety Program faculty provided assistance with survey data collection and evaluation and development of a training seminar to familiarize safety officers with results of the study.

Four Railroads made their properties available to the study committee for research:

Burlington Northern Railroad	Illinois Central Gulf Railroad
Duluth Missabe and Iron Range Railway	Southern Railway System

Their participation and support were largely responsible for the success of the study.

B.1.2.3 History – Need for Study

The goal of this study was to discover methods of improving safety program effectiveness in the railroad industry. When the need for such a study began to be apparent in 1975, the railroad industry, along with a number of other industries in the United States, was experiencing a continuing increase in the number of employee injuries. The recently implemented OSHA Law was having little or no effect in reducing accidents, even with the expenditure of large sums of money by the government and industry to achieve compliance. Clearly, the industry's safety efforts were not as effective as they once had been.

One reason for this lack of progress, in the view of many safety officers, was a tendency to cling to certain practices and beliefs about safety, which had little basis in fact. Research in the areas of psychology and sociology had clearly defined the role human behavior and motivation play in the workplace but few companies were applying the findings in their safety programs.

The OSHA Law, by requiring industry to pursue an "engineering approach" to the prevention of accidents, tended to reinforce the traditional practices in industry. A continued increase in the number of industrial accidents nationwide was ample proof that this somewhat simplistic approach of attempting to create a workplace, in which "a worker could not conceivably have an accident," was not effective. In fact, many industrial safety programs, in attempting to comply with OSHA requirements, seemed to be pursuing programming which was ineffective, or even counterproductive, from the standpoint of motivating safe worker behavior (see Figure B.1).

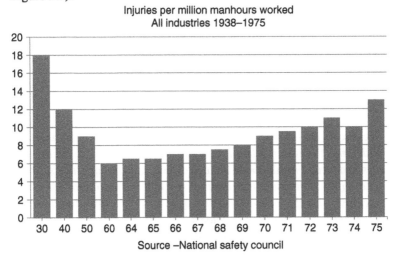

Figure B.1 Summary of injuries per million man-hours worked in Industry showing continued rise despite implementation of the OSHAct In 1970. Source: Based on National safety council.

In their landmark study of the impact of OSHA on workers in the chemical, aerospace, and textile industries, Northrup, Rowan, and Perry of the Wharton School, University of Pennsylvania, concluded:

> *The rather substantial changes in corporate commitments and federal requirements in the safety and health field resulting from the law have not been matched by parallel changes in the attitudes and behavior of most workers. Many workers, particularly older, long service workers, May find themselves forced to accept more safety and health protection on the job than they feel is necessary or desirable.*
>
> *There is ample evidence that regulation has fostered increased attention to safety and health issues and problems in all three industries examined here. There is little evidence, however, that this added attention has yet produced or is soon likely to produce a substantial improvement in the safety and health of most workers. This lack of significant impact exists because the added attention to safety and health generality has been focused primarily on the requirements of regulations instead of on the reduction of risks, in large part because the regulatory system focuses more on means than ends in the control of potential hazards.*

To reverse a similar trend in the rail industry, the AAR Safety Section's Steering Committee proposed to AAR management and to the FRA Office of Rail Safety Research that a joint study be conducted to examine present railroad safety programs, determine and quantify (if possible) the elements or activities which are associated with good safety performance, test elements which show promise in a railroad environment, and to make study results available to the industry for broader application.

In a paper entitled *Background and Needs of the Railroad Industry, the Steering* Committee summed up the need for this study:

> *In recent years, the railroad industry has been subjected to a great deal of public criticism in the area of safety. Pressure to reduce accidents has led to many studies and further regulation of isolated operating factors affecting safety.*
>
> *From a Safety Standpoint, the results of much of this regulation have been inconclusive. Tangible benefits in terms of reduced accidents and declining numbers of personal injuries still elude us.*
>
> *Individual member roads have already developed innovative programs to deal with some of the human factors problems in safety. An objective study of the*

relative effectiveness of these approaches and a determination of other pro-
grams which should be developed is overdue.

The AAR Safety Section representing 104 safety officers and 96% of the work-
force of the railroads of the United States and Canada, is the logical organi-
zation to make this study.

Allocation of a railroad's resources for safety purposes should be done in expec-
tation of success in controlling the perceived problems and returns to the com-
pany in terms of accident redactions, more efficient operations, and improved
employee relations. Perhaps the most useful information from the standpoint
of railroad management will be the criteria developed through the study to
evaluate program effectiveness.

Their recommendation was accepted and development of the study began in
1979.

B.1.2.4 Three Management Approaches to Safety Programming

The study group began its work with a search of literature and reports in the safety
field. It appeared that much of the safety programming currently being utilized
in industry could be placed in one of three categories, or approaches, based on
management methods and views of human behavior (Figure B.2).

The first two program approaches, were dubbed the "traditional" and "pro-
cedural/engineered" approaches. Seemingly, based on a concept of employees
as careless and in need of constant motivation, the first approach attempts to
motivate safe performance by enforcement of rules, policies and instructions, and
punishment for noncompliance. The latter attempts to motivate safe performance
through "engineering-out" hazards, developing written rules and procedures
for all operations, and using strong administrative action to motivate employee
compliance.

B.1.2.5 Philosophies Underlying Three Approaches to Safety Programming

The third approach, practiced in far fewer programs, could be called a
"behavior-based" approach. This approach characterizes employees as mature
human beings, largely motivated to work safely through behavior-based management systems. These programs encourage supervisors to deal with employees
as individuals, involving them in the safety program and in correcting systems
which produce unsafe employee behavior.

A review of the categories of activities included in these three types of programs
led to the conclusion that they have many elements in common. They differ, however, in the methods and the emphasis which managements place on the way

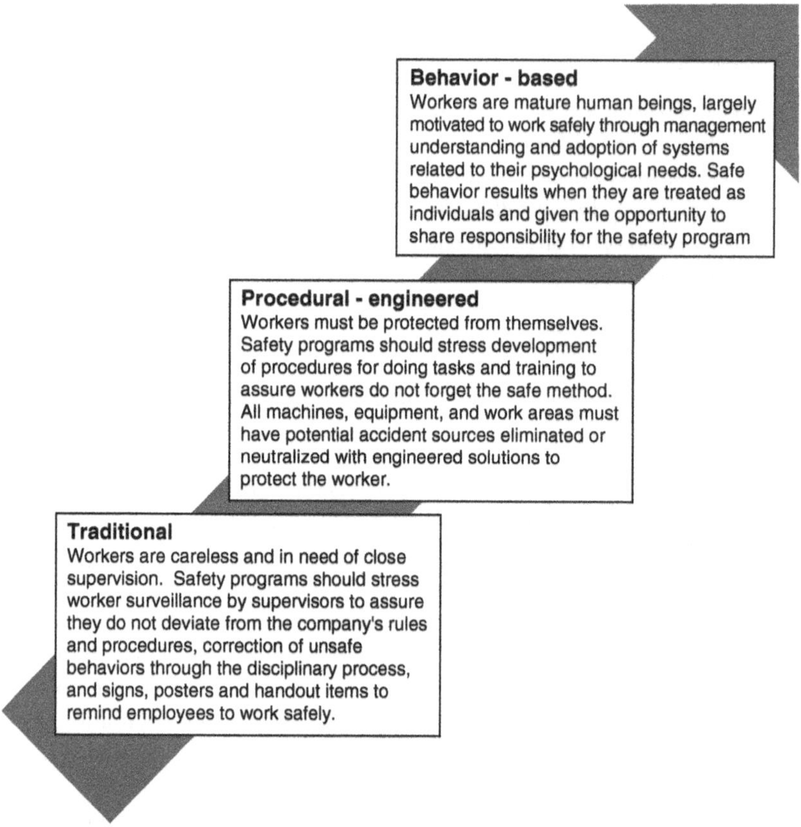

Behavior - based
Workers are mature human beings, largely motivated to work safely through management understanding and adoption of systems related to their psychological needs. Safe behavior results when they are treated as individuals and given the opportunity to share responsibility for the safety program

Procedural - engineered
Workers must be protected from themselves. Safety programs should stress development of procedures for doing tasks and training to assure workers do not forget the safe method. All machines, equipment, and work areas must have potential accident sources eliminated or neutralized with engineered solutions to protect the worker.

Traditional
Workers are careless and in need of close supervision. Safety programs should stress worker surveillance by supervisors to assure they do not deviate from the company's rules and procedures, correction of unsafe behaviors through the disciplinary process, and signs, posters and handout items to remind employees to work safely.

Figure B.2 Differences in approaches to safety programming. In practice, most safety programs contain elements of more than one approach. Assumption tested was that movement by a company toward a behavior-based approach would enhance safety program effectiveness.

activities are to be performed. For instance, both the "procedural/engineered" and the "behavior-based" may have a program for investigation of accidents. The first investigates to find which individuals were at fault and to limit liability. The latter investigates to find the flaws in the management system which failed to prevent the accident.

A 20 category summary, developed by the AAR Study Group (Table B.1) illustrates these differences in program emphasis and content.

B.1.2.6 Development of the Study Format

The pilot study was administered by the Ballistics Research Laboratory at Aberdeen Proving Grounds. This team of scientists had extensive experience

Table B.1 Twenty category summary of safety program differences.

Category	"Behavior-based" programs	"Procedural/engineered" programs	"Traditional" programs
Communications **Good programs have open lines of communication on safety matters. Employees have an active interest in matters related to the program.**	Information on safety matters is communicated to all levels. Included are progress toward goals, unit standings, and safe work procedures and rules.	Information on safety limited. Rules and procedures reviews are regularly conducted. Employees have difficulty relating to company safety goals.	Management does not make conscious attempt to inform with respect to safety record. Rules and procedures reviews are infrequent.
Training **Good programs provide training for employees in the safety aspects of their jobs. Employees feel training and method of instruction are adequate.**	Training programs follow guidelines requiring attention to content and context of message, quality of instruction, and follow-up to determine trainee understanding.	Training programs in safety matters are required. Little attention to quality of instruction or follow-up.	Training seldom conducted. Employees expected to learn safe procedures by reading rulebooks.
Attitudes toward safety **Good programs are characterized by positive attitudes toward management safety efforts and support for safe practices at all levels.**	Management aware of and works to eliminate causes of negative attitudes affecting safety at all levels.	Unaware of causes of negative attitudes in many cases. Attempts made to influence attitudes through campaigns and pronouncements.	Little concern for effects policies, instructions, and working conditions have on attitudes. Attempt to influence with "gimmicks" and "handouts."
Safety meetings **Good programs require regular, relevant safety meetings. Employees perceive them as favorably affecting safety performance.**	Employees involved in planning and conducting regularly held meetings. Records of proceedings kept and follow-up files maintained.	Supervisors responsible for conducting meetings on a regular basis. Often follow set pattern of management approved topics.	Meetings held irregularly as time permits. Often become "gripe sessions" related to unresolved safety problems.

Table B.1 (Continued)

Category	"Behavior-based" programs	"Procedural/engineered" programs	"Traditional" programs
Support for safety **In a good program, the whole organization is seen as working together to create a safe work environment.**	Management and employee support are present and visible. Safety problems not an issue in labor relations.	Management voices support but is not always perceived as following through in correction of safety problems.	Little evidence of management support and interest. Unions attempt to use grievance process to resolve safety concerns.
Inspections **Good programs require regular inspections of all areas with particular attention to high-hazard operations. Hazards are recorded and corrected.**	All aspects of operations, including work practices, covered by inspections. Employees frequently involved in inspections and follow-up on recorded hazards.	Supervisors perform required inspections. Focus is primarily on conditions and equipment rather than on employee work practices.	Supervisors inspect as time permits. Formal record seldom made of unsafe conditions.
Motivational programs **Good programs provide information on "off" as well as "on-the-job" safety matters to heighten employee interest in accident prevention.**	Employees participate in and view campaigns and programs as sincere and worthwhile expressions of management concern for their safety.	Motivational campaigns conducted regularly. Employees not involved and seldom interested.	Programs feature "gimmicks" and "giveaways." Seen by employees as attempts by management to "buy safety."
Rules violation – discipline **In good programs, the company is perceived to be taking a fair approach to handling of rules infractions.**	Emphasis on self-discipline. Firm, fair handling of employees who break rules with discipline as last resort.	Enforcement and investigation of violations emphasized. Supervision often overlooks infractions to avoid involvement in lengthy formal process.	Rules violations severely dealt with as a lesson to others. Employees look to union for protection from punishment.

Safety regulations In a good program, employees see safety rules and regulations as both necessary and adequate.	Company involves employees in discussions of rules and safe practices. Seen by employees as in compliance with government regulations.	Company insists employees "go by the book." Rules are developed and imposed by management. Employees see operation as "over-regulated."	Company requires strict rules compliance but sometimes seen by employees as ignoring its own rules. Lip service paid to federal rules compliance.
Accident investigation In a good program, focus of investigation is on understanding the cause and prevention of reoccurrence. Process is open and non-threatening.	Focus of investigation is on discovering and eliminating the underlying human and management systems contributing to the accident.	Investigation attempts to find primary cause of accident and fix responsibility. Underlying causes not dealt with.	Investigation conducted to fix responsibility for accident and assess discipline. Cover-up of information often occurs.
Recognition for good safety performance Good safety programs provide recognition for good safety performance at all levels of the organization.	Positive reinforcement of good behavior. Recognition for good safety performance in various ways.	Group incentives which are tied to statistics are used.	Recognition negative. Only employee mistakes receive attention.
Involvement of employees Good programs seek employee involvement in safety matters to promote identification.	Employees involved in safety policy and practices through safety committee activities, one-on-one consultation and individual safety-related assignments.	Little individual employee involvement. Management conducted safety meetings. Employees consulted infrequently on safety issues.	Program and rules imposed from above.
Safety climate In a good program, management's actions create a climate which encourages adoption of safe attitudes and work practices.	Management aware of requisites for creation of safe work climate and supports supervisor efforts in this regard.	Encouragement to adopt safe work practices hampered by perception of indifference to issues related to safety.	Management perceived as stressing production at the expense of safety.

Table B.1 (Continued)

Category	"Behavior-based" programs	"Procedural/engineered" programs	"Traditional" programs
Management credibility In a good program, management is seen as wanting a safe operation and willing to work for it.	Employees perceive high management interest and involvement in all phases of safety program.	Management voices support but seldom seen as interested or involved.	Management perceived as violating its own rules and standards for safety.
Quality of supervision In a good program, employees view supervisors as performing safety-related duties in a competent manner.	Supervisors deal with employees as individuals and encourage self-discipline. Good performance is recognized.	Supervisors attempt to achieve good safety performance through incentives and discipline. Employees dealt with as group rather than as individuals.	Supervisors regard employees as needing continual threats and prodding to achieve good safety performance.
Alcohol and drug abuse Good safety programs have effective means of dealing with alcohol and drug abuse problems.	Employees with problems dealt with promptly and effectively. Workforce educated to problems and supports company EAP program.	EAP program makes counseling available to employees. Referral largely through disciplinary process.	No EAP program. Problems ignored until discharge becomes necessary.
Employment practices Good programs are supported by employment practices which select, orient, and train new employees in accordance with job requirements.	Physical examinations, mental ability, and skills are basis for assignment to job. Supervisors or safe employees provide specific job orientation.	Employees hired on basis of screening interview applying set standards. General orientation to work followed by assignment to work with older employee.	Employees hired without regard to qualifications for job. Orientation to work superficial.

Supervisor training In good programs, employees perceive supervisors as being well trained and able to handle problems related to safety.	Supervisors are educated to human behavioral factors associated with safety and methods for conducting good safety programs in their areas.	Supervisors receive general training in management techniques.	Supervisors receive little or no training related to safety or management
Goals for safety performance Good programs involve entire organization in setting realistic goals and provide continual feedback on progress toward their attainment.	Employees encouraged to participate in goal-setting process and receive regular progress reports. Goals deal with problem resolution rather than numbers.	Goals for safety imposed from top. Typically involve achievement of a set number without regard to method.	No set goals for safety performance.
Hazards correction Good safety programs have established system for correcting reported hazards which is understood and supported by all levels of organization.	System provides for recording, assignment of responsibility for correction and feedback to person reporting hazard.	Supervisors responsible for correction and detection without established system.	Responsibility for reporting and correction hazards vague. Sometimes climate not conducive to reporting.

with safety through their involvement in the development of safety systems for weapons and space programs.

The Aberdeen Group proposed a study based on a "systems approach" to management of safety activities – an approach with which they were familiar and which they believed had worked well in the past.

The AAR Study Group, noting the need to go beyond present "traditional" and "procedural/engineered" models explained its concept of learning more about the effects of "behavior-based" safety activities. Conceding that few such program activities then existed, Dr. Dan Petersen pointed out the need to remedy the situation through testing of promising "behavioral" approaches to determine their effectiveness.

In the end, both approaches became a part of the study. Dr. Ed Baicy and his staff at Aberdeen Proving Grounds would evaluate the effectiveness of the "procedural/engineered" approach while the AAR Study Group, chaired by Chuck Bailey, evaluated the "human-behavioral" approach recommended by Dr. Dan Petersen and others.

B.1.2.7 Assumptions to be Tested

To provide direction for its efforts, the AAR Study Group listed five assumptions related to human behavioral factors and railroad safety programs which it felt should be tested:

1. *On every railroad there are a number of like "factors" present in the work environment – such things as fitness and ability to perform duty, management systems, condition of property and equipment, pressure, etc. – which combine in random fashion to produce accidents.*
2. *Railroad Safety programs consist of those systems and activities which have been established to influence these factors in a manner which will prevent accidents.*
3. *The best measure of quality of performance for a safety program is the perception of its effectiveness by all levels in the organization.*
4. *A great potential for improvement of safety program activities lies in better understanding of the work force and the way management systems affect it.*
5. *Railroads which have enjoyed the greatest success in reducing accidents and injuries are those which have developed a realistic approach to dealing with the human factor.*

B.1.2.8 Safety Program Activities Survey

To develop a baseline for the study, an extensive survey of railroad safety program activities was completed by each of the 18 members of the AAR Safety Section Steering Committee. These 18 railroads collectively represented more than 85% of the employment in the railroad industry at that time.

A detailed questionnaire was completed by the Chief Safety Officer and staff on each railroad. The questions were designed to determine the presence or absence of activities popularly thought to influence program effectiveness and to trace organization responsibility, staffing, program content, effectiveness measures, and perceived quality of performance. The information gained was then fed into a computer to be compared with other data measuring safety program effectiveness (Section B.1.2.11).

B.1.2.9 Involvement of Top Railroad Safety Officers

One of the strengths of the study was the continued involvement of the entire membership of the AAR Safety Section Steering Committee. This involvement of experienced safety officers provided the AAR Study Group with a valuable "real-world" focus for their efforts.

At frequent intervals during the nine-year course of the project, the study group provided presentations and updates on the study and solicited input in developing and testing measurements.

Steering Committee members were also involved in a three-day simulation/reaction exercise designed to test the usefulness of study data formats in analyzing safety program weaknesses and developing behavior-based solutions.

This attention to involvement throughout the period of the study kept the focus of the project on development of practical programs and measurements which would be of benefit to the industry. The study group was well aware that the usefulness of the final product to the railroad industry would depend largely on the understanding and actions of this Steering Committee. Association of American Railroads Safety Section Steering Committee members and Aberdeen Study Group meeting at the University of Minnesota – Duluth for a week-long information/reaction seminar which kicked off the study, shown in Figure B.3.

B.1.2.10 Pilot Survey – Railroads I and II

As a follow-up to the general safety program activities survey, to determine more precisely the factors involved in good safety performance, an in-depth survey of safety practices on two railroads was conducted. Railroad I had been a leader in safety performance for a number of years. Railroad II was near the bottom on the 18 railroad group in performance.

Questions relating to both the "procedural/engineered" and "behavior-based" approaches to safety, outlined previously, were included in the survey.

Five levels of each railroad's organization, from executives to hourly rated employees, were surveyed using questionnaires in booklet form, specifically designed for each level, but including some questions common to all. This permitted tracking of responses to common questions from level to level to determine where "gaps" in communications might be affecting safety performance.

Figure B.3 Initial study group photo.

The questionnaires were administered to a stratified random sample of managers and employees at five levels in the organization, chosen to provide statistical confidence of 95 + % at each level.

Safety officers on the two study railroads received training in administering the survey and answering questions from those being asked to participate. Every person surveyed had to be approached, invited to participate, and provided with a place where they would have privacy while completing it. Participants were given an addressed, stamped envelope in which to mail the completed questionnaire to the National Space Technologies Laboratory. There, all data were key-punched and processed by Computer Sciences Corporation's staff. This procedure generated a response of better than 97% of the questionnaires distributed on each property.

Numbers on each railroad participating in survey.

Level	RR I	RR II
Executive	3	10
Top management	21	47
Middle management	107	47
1st line supervisor	62	70
Employee	476	566
Total	669	712

It became evident that the methods used had produced a unique body of information for the Aberdeen and AAR Study Groups to analyze.

1. The 97% voluntary mail return rate is probably unprecedented for studies of this type.
2. No previous safety survey had ever attempted to measure responses from five organization levels in all functional areas. The analysis program also allowed evaluation of responses from specific functional groups within the organization.
3. It was evident, based on patterns of response, that differences between the two railroad safety programs had been identified.

Dan Petersen leads a discussion on human behavioral factors affecting safety at simulation exercise in Chicago in which 30 top safety officers used study formats to analyze safety program problems and devise solutions (Figure B.4).

B.1.2.11   AAR Study Group Analysis
The AAR Study Group worked primarily with responses to questions which indicated significant differences between the two railroads. A group of experts in railroad safety analyzed each response. Questions which provoked ambiguous responses due to wording or faulty premises were discarded.

Questions to which the percentage of positive response was substantially the same on both railroads (and, therefore, not indicative of any difference between the two) were also discarded. In this group were many of the questions which

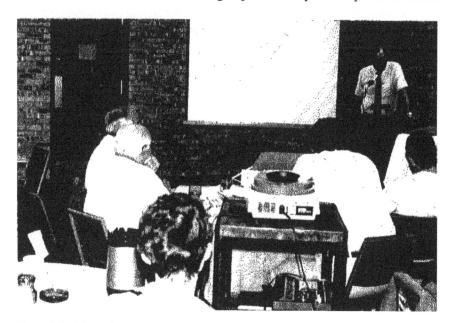

Figure B.4   Discussion group on human behavioral factors.

were based on the "procedural/engineered" approach. The remaining questions were subjected to further analysis. Possible communications, attitude, and morale problems were of particular interest to the study group.

Members working with the data were often surprised by the responses to individual questions. For instance, the relatively high degree of positive response from hourly rated employees belied the belief in some quarters that employees are generally apathetic or unconcerned about safety.

The AAR Study Group's approach to analyzing the data was based on the perception that the over-riding factor in safety program effectiveness is the ability of a railroad's management to understand and deal with those aspects of human motivation and behavior which lead to accidents. The AAR Study Group began by classifying questions with significant responses into management system categories.

Analysis of the data led to development of a "profile" for each railroad's safety efforts based on responses to survey questions (see Appendix 11). The response to questions on Railroad I – the railroad with the best performance as measured by other program effectiveness criteria – was significantly more positive overall than that for Railroad II. It was apparent that the survey had been able to accurately identify differences in safety program effectiveness on the two railroads.

B.1.2.12 Aberdeen Study Group Analysis

While the AAR Study Group was testing the human factor approach to safety programming, the Aberdeen Study Group attempted to find correlations between "procedural/engineered" oriented safety program elements and various safety performance indicators developed from data supplied by the railroad industry. Data from the Safety Program Activities Survey completed by 18 Class I railroads (Section B.1.2.8) was used to establish scores for each railroad in 12 subject areas.

1. Safety program content
2. Equipment and facilities resources
3. Monetary resources
4. Reviews, audits and inspections
5. Procedures development, review and modifications
6. Corrective actions
7. Accident reporting and analysis
8. Safety training
9. Motivational procedures
10. Hazard control technology
11. Safety authority
12. Program documentation

The Aberdeen Group's hypothesis was that high scores in these areas would correlate generally with lower accident and injury rates. Instead, they found little correlation with these "procedural/engineered" factors. Their report of findings stated, in part:

> *It was an unexpected result of this study that so little correlation was found to exist between actual safety performance and safety survey scores. The overall survey score has almost no correlation with train accident rates and cost indicators and is somewhat counter-indicative with respect to personal injury rates. The only two categories which correlated consistently and properly with accident rates were monetary resources and hazards control technology. Two categories, equipment and facilities resources and reviews, audits and inspections, had counter-intuitive correlations.*

This conclusion coincided with the findings of the AAR Study Group which found that responses to most pilot survey questions based on these 12 criteria did not distinguish any significant differences between the two railroads studied.

Taken together, the results achieved by the AAR and Aberdeen Study Groups seemed to be saying:

1. No two railroad safety programs were alike in content, organization, or emphasis.
2. The effectiveness of safety programs cannot be measured by the more traditional "procedural/engineered" criteria popularly thought to be factors in successful programs.
3. A better measure of safety program effectiveness is the response from the entire organization to questions about the quality of the management systems which have an effect on human behavior relating to safety.
4. The most successful safety programs are those which recognize and deal effectively with employee and supervisor behavior and attitudes which affect safety.

Chairman Chuck Bailey (right) and Jack Southworth of the AAR, using survey response data charts to place pilot survey questions in categories (Figure B.5).

B.1.2.13 Further Refinement of the Survey Process

The conclusion of the pilot survey phase of the project left many unanswered questions. Clearly, it had been proven that many of the things commonly thought to influence safety were having little or no effect. In its report, the AAR Study Group cautioned that results could only be considered preliminary and recommended that they be used to develop an instrument and process which could be initiated by companies wishing to evaluate the overall effectiveness of their safety efforts.

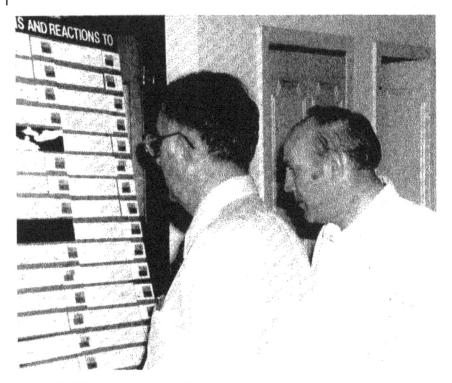

Figure B.5 Pilot survey questions discussion.

Criteria were established by the AAR Study Group to guide development and evaluation of a second "verification" survey which they proposed be conducted on two additional railroads. This second survey should:

1. Evaluate the organization's perception of management systems which affect safety performance.
2. Ask the same questions of managers and employees at five different levels in the organization.
3. Be easily and economically administered, analyzed, and evaluated without using a main-frame computer.
4. Allow identification and comparisons of specific departments and divisions while maintaining the respondent's anonymity.
5. Provide managers with data in a format which will allow definitive comparisons and decision-making.

B.1.2.14 Survey Verification Study – Railroads III and IV

With the approval of the AAR Steering Committee and the FRA, the study group recruited two additional railroads on which to test the revised survey plan. The top safety officers of these roads and their staffs worked with the study group in development and planning of a survey to fulfill the requirements of the established criteria.

The survey was administered to a stratified random sample of employees and management on both railroads.

A total of 1825 completed questionnaires were received from the two railroads, allowing a 95 + % confidence level with respect to inferences concerning an entire railroad and a confidence level of 90 + % at the region and division levels.

Numbers on each railroad participating in survey.

Level	RRIII	RRIV
Executive	14	1
Top management	21	6
Middle management	184	16
1st line supervisor	372	23
Employee	878	165
Unspecified	128	17
Total	1597	228

B.1.2.15 Description of Analysis Program

Questions which had elicited a statistically significant response on the pilot survey were assigned to 20 categories, each of which defined a specific management system affecting safety. Each question was assigned to one or more categories for the purpose of identifying response patterns.

Analysis programs were written in BASIC for the IBM PC-XT. They allowed the user to specify search criteria based on any combination of responses by railroad, or organizational unit within a railroad. The program also permitted the creation (either on an ad hoc basis or permanently) of additional categories or grouping of questions which appeared to be meaningful. Two output formats were available:

1. A display of all question categories and response values for a specified combination of search criteria. It included an optional listing of response totals for each individual question. One aggregated report was created for each inquiry.
2. A display of multiple reports for specific organization units in three question categories without detailed individual question counts.

Results from either report format could then be tested for statistical significance.

Data were then converted to easily read bar charts, which allowed comparisons between departments, divisions, regions, and railroads for each of the 20 categories. Other graphic presentations of differences in response between levels in

Figure B.6 Minnesota graduate students.

the organization from top executives to employees and areas of high and low positive response were also developed. This information was then presented to top management on both railroads to aid in analysis of safety program effectiveness (Appendices 1–5).

Minnesota graduate students John O'Brien and Linda Malm demonstrate use of SCANTRON to read data from questionnaires directly into computer for AAR Study Group. Left to right: Hatchard, Bain, Underwood, Kuhlman, Thomas (FRA), Petersen and Andrews are shown in Figure B.6.

B.1.2.16 Analysis and Use of Survey Data by Managements

Comparison of data from the survey with other indicators of safety performance confirmed that units with the highest positive response to survey questions were generally those with the best performance as measured by other indicators of safety performance.

The analyses of survey data provided by the AAR Study Group proved to be extremely useful to managements on the participating railroads. For instance, on one railroad, the survey clearly identified the weaker management systems which were affecting safety performance such as:

> *"Recognition for Good Safety Performance"* – with a low 48% favorable response from hourly-rated employees.
> *"Inspections"* – with only 51% favorable response.

"Supervisor T raining" – 49% favorable response.
"Quality of Supervision" – 50% favorable response.

This cluster clearly indicated a major problem in supervisory performance and led to training of supervisors in positive reinforcement and employee assessment techniques on selected divisions during the second phase of the study.

On this same railroad, the widest difference in perception of program effectiveness between hourly rated employees and management was in these same four categories. Not only did they have a serious problem, but they were largely unaware the problem even existed.

One of the regions of one study railroad was consistently lower in positive hourly rated response than all of the other regions in nearly every category. Management on this region clearly had a severe credibility problem with employees in nearly every phase of their safety efforts (Appendix 3).

A variation in regional response of 24% for one category – Appendix 3, "Accident Investigation" – pointed to a possible problem in this regard. At the lowest positive response level of 40% for "Quality of Supervision" on this same chart, 6 out of 10 employees were questioning the competence with which their supervisors performed duties related to safety.

Managements on the study railroads were also able to compare survey response for their railroad with that of other roads to determine whether a low score was indicative of a local railroad or an industry-wide problem. Further study of the reasons for these differences may lead to a determination that one railroad has developed a superior method for handling that particular area which should be adopted by others (Appendix 6–7).

The AAR Study Group has been monitoring the effect of presentation of data to managements on the study of railroads with the following results:

1. One of the railroads is using the survey results to develop performance objectives for individual managers and safety committees. The top executive has decided to personally review the findings with safety committees which did not meet their previous year objectives.
2. Safety officers are making use of the data for their respective responsibility areas to call to managers' attention to possible problems which require corrective action. Their involvement in the survey process provided them with a more balanced picture of the management activities which affect safety performance.
3. Because the survey provides a "picture" of the attitudes and beliefs of the organization at a specific point in time, the database can be compared with data from future surveys to determine the effects of new safety initiatives. One of the railroads is currently resurveying in selected areas to determine the effects of the training program conducted as a part of the study.

4. Other railroads using the survey to measure safety program effectiveness will have access to a large and complete database with which to compare their data. The AAR Study Group recommends making a "survey package" of software and instructions available to assist member roads in this regard.

Data from the surveys on all four railroads have been compiled and appear in Appendices 6–7.

B.1.2.17   Testing a Human Behavioral Factors Approach

Conclusion of the survey phase of the project left the AAR Study Group with many unanswered questions. Clearly, it had been shown that many of the things commonly believed to influence safety programs were having little or no effect. All data seemed to point to using human behavioral principles to design programs to reduce unsafe employee behavior.

A number of presentations were made by the AAR Study Group at regular meetings of railroad safety officers. It was the general consensus, based on findings of the study, that further testing of the hypotheses related to the importance of human behavioral factors should be done. In discussions with Research and Development people at the Federal Railroad Administration, possibilities for a human factors demonstration project were examined.

Dr. Dan Petersen, the study group's expert on the behavioral aspects of safety, developed a method of testing the effect of a behavior-related change in the management system on the workforce.

Evaluation of behavior-related principles and a review of areas of low positive response from the survey data led to selection of the techniques of "positive reinforcement" and "individual assessment" to be tested. Firstline supervisors would be trained to use the techniques and management system would be established to encourage their use. Also, a system of measuring worker behavior to determine results of the training would be installed.

The four railroads which had participated in the safety program effectiveness survey also volunteered to make their facilities and employees available to the AAR Study Group for this behavior intervention experiment. The information on their safety programs already available would, it was felt, be of real benefit to the study group in the assessment of results.

B.1.2.18   Technique to Measure the Effects of the Experimental Program

It was clear that the traditional measures of safety performance – accidents and injuries – would not provide an accurate measure, in the short run, for the changes in practice being asked of the supervisors. The only way to accurately assess changes in behavior due to changed practices was through direct observation of workforce behavior.

To rule out many of the other variables in the work environment which might influence results, it would also be necessary to measure workforce behavior in comparable units where supervisors received no training.

On most railroads, a "division" is both a geographical and administrative entity with clearly defined boundaries and management responsibilities. It was decided to provide training to all firstline supervisors on one division on railroads I and II and measure the effects of the training by observing worker behavior on that division as well as on a division with similar characteristics where supervisors received no training.

Safety officers on both railroads, who would be making the behavior observations, were brought together to develop standard lists of unsafe practices. They then received training in the methods to be used in making observations. The need to observe all areas of each division's operations on a random basis was particularly stressed.

Before training of supervisors began, these observers made "base-line observations" of the workforce. Sufficient observations were made to assure statistical validity and reliability.

On one railroad, unsafe behavior was noted in 35% of the observations made on the division to receive training and 34% on the "control" division.

On the other railroad, 39% of the observations on the "training" division involved unsafe actions, and 31% on the "control" division.

Observers on two divisions reported their perceptions that the number and types of unsafe behaviors observed seemed to correlate with the division's accident experience.

B.1.2.19 Training Format – Railroads I and II

The training given firstline supervisors outlined a number of principles of human behavior which relate to safety performance as well as studies which have dealt with ways of eliminating undesired behaviors. The studies on use of positive reinforcement techniques to achieve improved performance were also discussed in detail. Supervisors were then asked to change one thing in their method of supervising their people. They were asked to adopt the practice of giving positive recognition to individual workers observed doing their jobs in a safe manner and to provide weekly reports of the number of such contacts made to their supervisors.

At the close of the training session, participants completed a questionnaire on which they were asked to give the number of positive contacts per week they had been making prior to receiving training. They were also asked questions on the content of the training program. These answers were then compared to the answers given on a pre-trained test of their knowledge. Improvements of 519% on one railroad and 769% on the other were noted, indicating that the principles taught were well understood.

Detailed results of these measures are shown in Appendix 8. In addition to the pre and post tests on content of the training (Hours 1–6: Positive Recognition Training Outline), supervisors were also asked to give their reaction to the training and to the concept of positive reinforcement (see Hour 1 and 2: Follow-up session outline; hour 3: **Follow-up session outline**). Reactions were favorable.

They were also asked to write down the number of safety-related activities they had completed in the week prior to training to get a picture of the level of activity as a baseline (see section C).

Appendix 9 contains an outline of the course material presented.

Realizing that training without support from top level managers is seldom effective, the study group held orientation meetings with the top executive management people on each railroad. Conducted by Dan Petersen, these meetings reviewed study findings to date and outlined the purpose and content of the training exercise. The need for support from the line organization for the changes in supervisory practice being asked of firstline supervisors was particularly stressed.

With executive approval and support, the study group scheduled meetings with management on the divisions where supervisors would be trained. A two-hour orientation session was held with the division superintendent and his staff. The content of the training program and changes in practices being asked of supervisors were outlined. Management's cooperation was also sought in establishing a system of reporting which would place emphasis on the use of the techniques being taught.

Each division adopted its own reporting system. Most involved a small card, turned in weekly, which listed the number of supervisor contacts on a day-to-day basis. It was made clear to the supervisors that they should not participate if they did not feel comfortable doing what was being asked.

B.1.2.20 Results of Positive Reinforcement – Railroads I and II

For three months, the superintendent monitored supervisory activities on the basis of reports of contacts being made. At the end of that period, a follow-up round of observations of workforce behavior was made. On both of the divisions on which supervisors had received training, the number of observations of unsafe behavior had dropped; from 35 to 20% on one railroad and from 39 to 23% on the other. On the divisions which had not received training, the number of unsafe behaviors remained nearly constant. Details are shown in section D of Appendix 8.

It was evident that the practice of positive reinforcement by supervisors was having a real effect on the actions of the workforce.

A second training session was conducted at the end of three months. In these sessions, supervisors were asked about their experiences in using the positive recognition techniques and whether they thought it was doing any good. They were then asked to try an "employee assessment" technique, in addition to the one they were already using, for the next three-month period (see Appendix 10).

Questionnaires completed during this second session indicated that supervisors had increased the number of "positive reinforcement" contacts over those reported at the first session by 56% on Railroad I and 48% on Railroad II. Responses also showed that on Railroad I, there was no reduction in other safety-related activities (meetings, inspections, training, discipline, etc.). On Railroad II, these other safety activities decreased by 34%.

Three months after the second training session was conducted, a final round of observations of workforce behavior was made. The results indicated that use of the employee assessment technique did not produce further significant reductions.

On Railroad I, which had the best safety record in terms of conventional measures going in to the study, unsafe behaviors were reduced by 41% on the division which received training and 11% on the control division. On Railroad II, unsafe behaviors were reduced by 49% on the training division and 3% on the control division. Slight improvements on the control divisions may possibly have been due to the presence of the observers.

Railroad II also experienced a significant improvement in terms of the conventional measures of safety performance in the years which followed. The General Manager – Safety and Training attributed much of this change to the interest in safety generated through involvement in the study.

B.1.2.21 Verification of Results on Railroads III and IV

This experiment in positive reinforcement was conducted again in 1987, on two additional railroads on a larger scale. In the first phase, only two divisions on each of two railroads had been used (one experimental group and one control group). In the second phase, eight divisions from the two additional railroads were used (four experimental and four control). The same format and measures previously described in phase I were used for this experiment.

Pre and post tests to determine knowledge gained through training showed a gain of 560% on one railroad and 298% on the other (see Appendix 8).

Reaction of the participants to the training given and to the concept of positive reinforcement as a supervisory technique was also favorable at all locations. Reaction to use of positive reinforcement, though, was somewhat lower on Railroad IV.

Supervisory performance, as measured by daily safety activities, showed increases on all divisions where training took place.

One further analysis determined whether an increase in the requested activity (positive reinforcement, contacts, and observations) resulted in a reduction in other safety-related activities. Supervisors on three divisions reported increases in other safety activities while those on two reduced their other safety activities.

B.1.2.22 Reductions in Unsafe Behaviors

As in Phase I, the primary measure of positive reinforcement effectiveness used was reduction of unsafe behaviors as measured through behavior observations made by safety officers. On Railroad III, results were influenced by many unforeseen factors and events which made precise measurement of the effects of positive reinforcement training difficult.

The experimental divisions on Railroad III achieved an 11% improvement in the reduction of unsafe behaviors while control divisions recorded an 8% improvement. While considerably below reductions obtained in previous experiments in positive reinforcement (in both railroad and nonrailroad applications), the study group believes the results are what should be expected, given the following circumstances:

- The poor performance of experimental Division M in the first three months (without Division M, the three-month improvement for the experimental divisions was 26%).
- The poor performance of experimental Division W in the second three months, (if this division had maintained its three-month performance level, the six-month improvement for Railroad ill would have been over 17%. A heavy workload placed on supervisors due to a large increase in business without a corresponding increase in workforce and supervision undoubtedly had an effect in this instance).

Also contributing to limited improvement was the fact that supervisors on two experimental divisions (Division M, to a small degree, and Division A, to a rather larger degree) had been previously taught positive reinforcement techniques and, in the case of Division A, had been required to use them for over a year. Little improvement, beyond what would accrue from "brush-up" training would be expected in this situation.

On Railroad IV, the experimental division recorded a 25% improvement for the first three-month period. This improved to 28% at six months. The control group recorded a 7% improvement at three months which increased to 16% at six months. The control division improvement for six months is probably due to the transfer of employees from the experimental to the control division, due to seasonal workforce changes, just prior to the last round of behavior observations.

B.1.2.23 Summary of Positive Reinforcement Experimental Results

The purpose of this phase was to determine whether behaviors contributing to accidents could be reduced through the application of one technique – positive reinforcement.

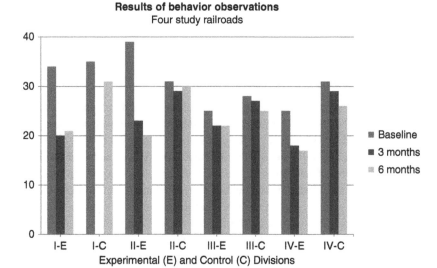

Results of behavior observations
Four study railroads

Figure B.7 The percentage of unsafe acts observed in the "baseline" period before training began ranged anywhere from 25 to 38% (left bar). The chart graphically illustrates the effect produced by positive reinforcement training on the experimental (E) divisions. Little change can be seen after six months on the control (C) divisions whose supervisors received no training.

Where it had been tried in a variety of other industrial experiments, this technique had been successful and resulted in some positive behavior change in the workforce.

With four railroads involved, this was perhaps the largest experiment ever conducted using this technique. A total of 12 divisions, 415 supervisors, and approximately 10 000 hourly employees were a part of the experiment.

The ability of positive reinforcement to enhance safety program effectiveness was clearly demonstrated. Figure B.7 depicts the results of positive reinforcement training done in connection with this study.

Impact of Study – Four Railroads

B.2 Railroad I

B.2.1 Background

At the time the study was conducted, Railroad I operated in excess of 10 000 miles of track with 20 000 employees and served 13 southern states east of the Mississippi. In the 1970s, it was one of the top-ranked railroads in the nation in terms of low accident rates. The safety department of 21 officers was given high visibility and support by top management. Management tended to use a strong procedural-discipline approach to achieve good safety performance.

B.2.2 Impact of Study

The process involved in instituting the study on this railroad and the follow-up emphasis on positive behavior reinforcement served as a catalyst in changing management thinking with respect to the supervisor's role in safety. One observer noted that the railroads' reliance upon a strict disciplinary approach had reached its limit as a motivator of safe performance and that what was learned through the study resulted in a more enlightened behavior-oriented approach to supervision. A follow-up survey of supervisory practices conducted four years after the positive reinforcement training found nearly all of those responding were still using the techniques and enthusiastic about the benefits of doing so. In the years following completion of the study on this road, safety performance, in terms of low accident rates, continued to be excellent. In the study group's view, an already good safety program had been enhanced by what was learned through the study.

B.3 Railroad II

B.3.1 Background

At the time the study was undertaken, Railroad II operated more than 9000 miles of track from the Great Lakes to the Gulf of Mexico, with some 16 600 employees. In the late 1970s, it had a poor safety performance in terms of casualty data. The safety staff consisted of a small headquarters group. Accident prevention activities were monitored by a general safety committee, just prior to the study, management made several changes to improve. Safety Staff reporting was changed from Personnel to the Operating Department and additional field safety officers were appointed. Line operating departments received safety budgets and safety performance became a part of the performance appraisal process for line officers. Remedial safety training for employees was also conducted.

B.3.2 Impact of Study

Institution of the study served as an effective backdrop for the safety policy changes being instituted. Perceptions of safety policies and procedures were solicited at all levels of the organization – from the president's office to the switch shanty. Promotional programs were decentralized and individual and unit safety achievements were recognized. A back injury prevention program, staffed by trained employees was conducted. Over a six-year span, from 1978 through 1984, casualties were reduced by 68%. The railroad also worked an unprecedented 2½ years without a fatality. In an article in a leading newspaper, Railroad II's General Manager – Safety was quoted, "I think one of the things we learned was that there is such a thing as 'positive behavior reinforcement'. Instead of just getting after the bad apples, like we used to, we learned to give credit to people who were doing a good job."

B.4 Railroad III

B.4.1 Background

Railroad III is a large western railroad which was formed through the merger of several roads. Throughout the study period, this railroad was engaged in reducing duplicate trackage, facilities, and responsibilities. One entire region was eliminated and several changes in territories of the remaining regions were made. At the time of its participation in the study, the railroad operated in excess of 25 500 miles of track and some 35 000 employees. Safety responsibility had been decentralized. Regional safety officers were responsible for development and implementation of region safety programs with support and consultation from a central staff group. These programs varied in approach from "behavior-based" to the more "traditional." Divisions from three separate regions were chosen to participate in the training phase of the study and each of these experienced changes which affected study outcomes.

B.4.2 Impact of Study

Implementation of the survey phase of the project served to focus management's attention on the whole range of management systems which were having an effect on safety performance. Results were first reviewed with executive management who requested that region management use them in developing new programs. Results of the training phase indicated that the majority of supervisors who received training used the positive reinforcement techniques with good results and intended to keep on using them.

B.5 Railroad IV

B.5.1 Background

Railroad IV is a wholly owned subsidiary of a major steel corporation. It operated 1000 miles of track with some 800 employees. For many years, this road has been a leader in development of innovative safety programs. Safety performance, in terms of casualty data, was excellent. Just prior to the study, the safety department on this road had been eliminated, with responsibility assigned to 13 management-employee safety committees reporting to an Executive Safety Committee.

B.5.2 Impact of Study

Results of the survey portion of the study were reviewed with the Executive Safety Committee. Each department head was asked to use the information to enhance present safety efforts. Individual performance objectives based on survey results are also being adopted. The top executive will also be reviewing findings of the survey with those safety committees who did not meet their previous year's objectives.

One-half of the supervisors on this railroad received positive reinforcement training. Many of these (85%) had been exposed to the concept in prior training. A significant reduction in unsafe behaviors was noted for employees whose supervisors received training and all supervisors responding to a survey after the project ended indicated they intend to continue to use the technique.

B.5.2.1 Longer Term Use of Positive Reinforcement

A follow-up survey of supervisors who received positive reinforcement training on five of the six experimental divisions was conducted by the AAR Study Group. A summary of the percent of positive responses to questions appears below:

Question	Divisions				
	I (%)	II (%)	III (%)	IV (%)	V (%)
Was the training received of value to you?	88	100	100	100	100
Did employee safety behavior improve?	75	33	42	56	77
Will you continue to apply the principles learned?	88	100	100	100	100
Should similar training be given to all supervisors?	100	87	83	89	92

Their response clearly indicated that the overwhelming majority of supervisors who received the training considered it valuable, felt others should also receive the training, and intended to keep on using positive reinforcement techniques themselves.

B.5.2.2 Study Conclusions and Outcomes

In reviewing the mass of data generated by this nine-year study, the MR Study Group was able to make the following statements with respect to current safety practices and use of the techniques developed to enhance safety program effectiveness.

1. The 12 most commonly used criteria in standard safety program audits are poor measures of program effectiveness.
2. Application of the 20 category survey techniques developed by this study provides a reliable measure of safety program effectiveness.
   a. The AAR survey effectively identifies the strengths and weaknesses of the 20 management system areas affecting safety performance.
   b. The AAR survey effectively identifies major discrepancies in perceptions of effectiveness of safety program elements between hourly rated employees and various levels of management.
   c. The AAR survey effectively measures improvement and deterioration of safety program effectiveness when administered periodically.
3. Use of positive reinforcement techniques will enhance safety program effectiveness.
4. Behavior observation, properly administered, is an effective tool for measuring the effects of applying specific management techniques in the railroad setting.

In addition to the conclusions stated above, the AAR Study Group summarized the specific products and outcomes of the study as follows:

1. An economical method of evaluating the effectiveness of a company's safety program has been developed and tested with positive results.
2. A behavioral technique to improve workforce safety performance has been adapted to the railroad environment and tested with positive results.
3. A software package for railroads wishing to implement the evaluation process has been developed and made available through the AAR.
4. An observation technique for measurement of the effect of training on work force behavior in the railroad setting has been developed and tested with positive results.
5. A curriculum which uses simulation to train those responsible for safety to use study concepts and data formats to analyze and enhance safety program effectiveness has been developed and tested by a group of 30 railroad safety officers with positive results.

The understanding created by the study has focused attention on analysis and methods of improvement which enhance safety program effectiveness. With continued study and wider application of the findings of this study, the AAR Study Group believes the potential for continued industry improvement is excellent.

B.5.2.3 A Final Word

Since this study began in 1979, many individuals and organizations have made valuable contributions to our study gGroup's efforts. We wish to especially recognize the contribution of Frank Kaylor, Assistant Vice President – Safety and Hazardous Materials of the Southern Railway System, one of the original members of the AAR Study Group whose untimely death prevented his sharing in the completion of this project.

Without the support given us by management on the four study railroads, we could not have conducted this study. It is our hope that the results achieved will balance out their contributions of time and facilities to conduct the project. We are also grateful for the vision and support of those officials in the Federal Railroad Administration who saw the potential for improvement in the understanding of human factors which this study provided.

We also wish to express significant thanks to the faculty of the University of Minnesota-Duluth for their help and support and for use of their facilities. Their blend of academic discipline with real-world problem-solving contributed greatly to the final product.

Appendix 1: Sample – Chart Used for Analysis on One of the Study Railroads

Railway total response by category

1. Low response by level
   (below 70% positive response)
   a. Top Mgt.
   b. Mid Mgt.
   c. Supers. –recognition for safety performance
   d. Employee – training, support for safety, recognition for safety performance inspections, management credibility, supervisor training, quality of supervision rules violations/discipline
2. Perceived strengths
   (above 80% positive response)
   Motivational programs, involvement of employees, goals for safety, accident investigation, safety meetings
3. Possible communications problems
   (More than 20 points difference between Top Mgt., and any other level)
   Inspections, support for safety, attitudes toward safety, management credibility, quality of supervision, training, communications

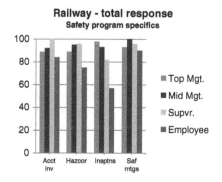

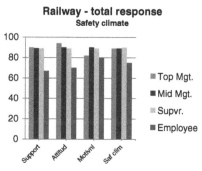

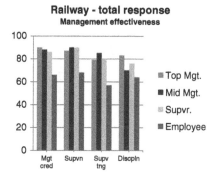

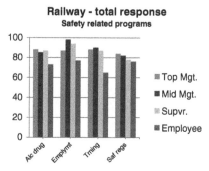

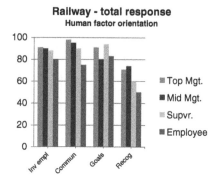

Appendix 2: Sample – Chart Used for Analysis on One of the Study Railroads

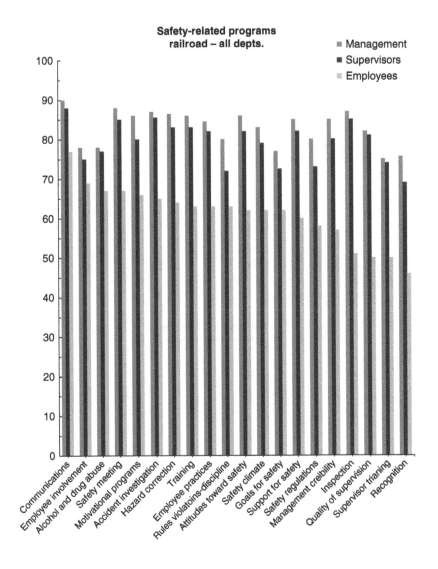

Appendix 3: Sample – Chart Used for Analysis on One of the Study Railroads

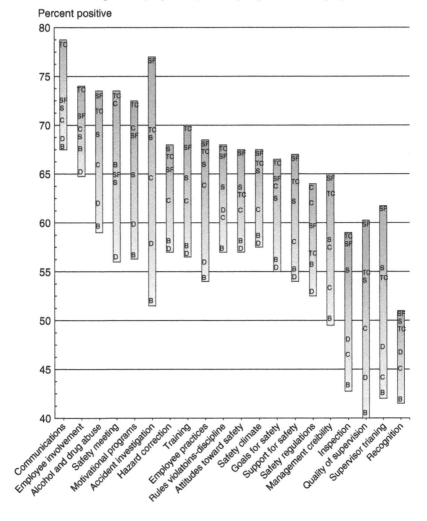

Range of employee response by region and category

Appendix 4: Sample – Chart Used for Analysis on One of the Study Railroads

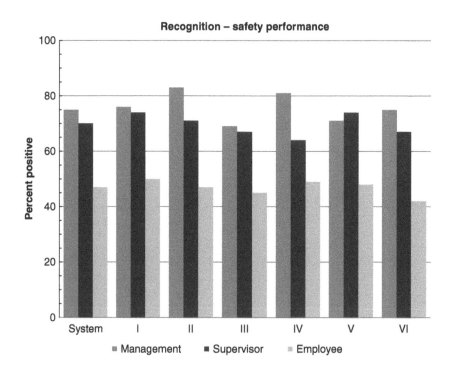

Recognition – safety performance

R, Recognition for good performance: good safety programs provide a means of recognizing good safety performance at all levels of the organization.

Questions:

26. Is safe work behavior recognized by supervisors?
39. Can first line supervisors reward employees for good safety performance?
47. Is safe work behavior recognized by your company?
59. Are safe workers picked to train new employees?
70. Is promotion to higher level jobs dependent upon good safety performance?

Appendix 5: Sample – Chart Used for Analysis on One of the Study Railroads

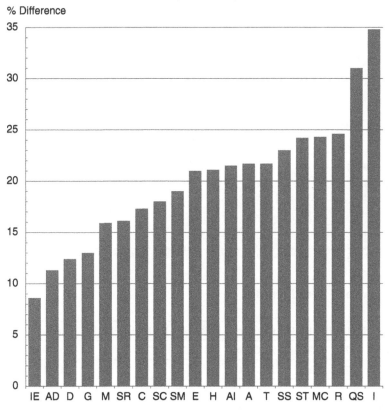

Safety survey response

Difference between management and scheduled positive responses

% Difference

AI	Accident investigation	H	Hazard correction	I	Inspection	SM	Safety meetings	IE	Employee involvement
G	Goals for safety	C	Communication	SS	Support for safety	A	Attitudes toward safety	M	Motivational programs
R	Recognition of performance	AD	Alcohol and drug abuse	SC	Safety climate	E	Employee practices	SR	Safety regulations
QS	Quality of supervision	D	Rules violation-discipline	T	Training	ST	Supervisor training	MC	Management credibility

Appendix 6: Total Response – 20 Categories – 4 Railroads

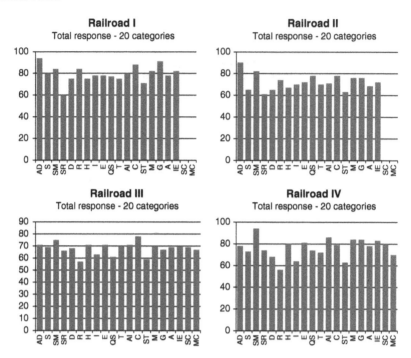

Comparison – 4 study roads

RR I	RR II	RR III	RR IV	Category	RR I	RR II	RR III	RR IV	Category
81	71	73	86	ACCT INV	83	75	72	84	MOTIVNL
77	66	72	80	HAZ COR			70	79	SAF CLIM
79	70	64	65	INSPTNS	94	90	71	77	ALC DRUG
83	81	75	93	SAF MTGS	79	72	71	81	EMPLMT
83	72	72	83	INV EMPL	76	70	71	72	TRNING
88	78	78	79	COMMUN	60	61	66	74	SAF REGS
92	75	67	84	GOALS	78	78	62	74	Accident Prone Handling Quality Supv*
83	74	57	56	RECOGN	73	64	60	65	SUPV TNG
81	65	69	73	SUPPORT	76	65	68	68	DISCPLN
79	68	70	77	ATTITUD			67	71	MGT CRED

Appendix 7: Comparison of Positive Responses by Category – 4 Railroads

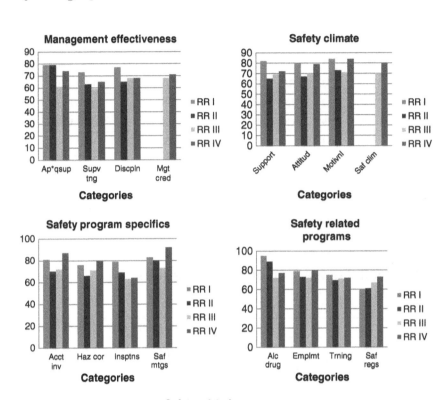

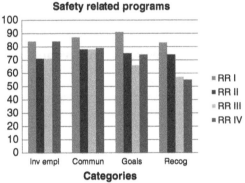

Comparison – 4 study roads

RR I	RR II	RR III	RR IV	Category	RR I	RR II	RR III	RR IV	Category
81	71	73	86	ACCT INV	83	75	72	84	MOTIVNL
77	66	72	80	HAZ COR			70	79	SAF CLIM
79	70	64	65	INSPTNS	94	90	71	77	ALC DRUG
83	81	75	93	SAF MTGS	79	72	71	81	EMPLMT
83	72	72	83	INV EMPL	76	70	71	72	TRNING
88	78	78	79	COMMUN	60	61	66	74	SAF REGS
92	75	67	84	GOALS	78	78	62	74	Accident Prone Handling Quality Supv*
83	74	57	56	RECOGN	73	64	60	65	SUPV TNG
81	65	69	73	SUPPORT	76	65	68	68	DISCPLN
79	68	70	77	ATTITUD			67	71	MGT CRED

Appendix 8: Comparison of Training Results – 4 Railroads

	RR I	RR II	RR III	RR IV	Total
A. Knowledge Gain as measured by pre- and post-test scores	519%	769%	560%	298%	537%
B. Reaction to course Average reaction to training session (scale 0–5)	4.38	4.23	4.21	4.1	4.23
Positive reaction to request to use technique (scale 0–5)	4.29	4.18	4.14	3.75	4.09

C. Change in performance reported by supervisors – safety related activities per day

Prior to training (scale 0–5)	3.20	4.40	2.93	1.86	3.10
Three months later (scale 0–5)	4.40	5.00	4.23	3.74	4.34
Improvement	40%	14%	44%	101%	40%
Positive reinforcement increase	56%	48%	62%	112%	70%
Increase in other activities	4%	−34%	24%	86%	20%

D. Behavior sampling

	RR I		RR II		RR III		RR IV		Total	
	Stdy	Ctrl	Stdy	Ctrl	Stdy	Ctrl	Stdy	Ctrl	Stdy	Ctrl
	(%)	(%)	(%)	(%)	(%)	(%)	(%)	(%)	(%)	(%)
Baseline	34	35	39	31	25	28	25	31	31	31
3 Months	20		23	29	22	27	19	29	21	28
Percent improvement	+43	34	+41	+6	+12	+1	25	+7	+32	+10
6 Months	21	31	20	30	22	25	18	26	20	28
Percent improvement	+40	+11	+49	+3	+11	+8	+28	+16	+35	+10

Appendix 9: Positive Recognition Training Outline

Hour 1–2 Introduction
Overview of the study Pre tests and questions
Introduction to the session
Why try a different approach to safety
Updating safety theory
• What we've always done vs. what we now know
• What an accident is, and what causes them
—Unsafe acts/conditions vs. human error
• What we can do to prevent accidents
—The 3 E's of safety vs. the 3 behavior changers
• What other companies and industries have found
Hour 3 Worker motivation and what we know about it
What motivation is
• A person doing whatever is necessary to satisfy his current needs
What needs are
• The major theories and concepts
—Needs of most workers
—Needs change
—Motivators and dissatisfiers
• What the research shows
Hour 4 Influencing worker behavior
Your three choices
• Motivation – the environment
• Attitude change
• Changing behavior

Motivation – the easiest way
- What influences workers usually
—The peer group
—You and your style of leading
—Your credibility (particularly in safety)
—Your attitude toward safety (what they see)
—How you measure and judge them
—What they think your priorities are
—The organizational climate

Hour 5 What you can do

Understand each worker use the motivators
Understand the group of workers you supervise

Hour 6 Using positive recognition

Follow-up session outline

Hour 1 Review of the session 1 concepts

Motivation concepts
The influences on worker behavior
Positive recognition

Hour 2 How you did in the last 3 months

Difficulties successes

Hour 3 Ensuring you understand each worker

Your task for the next three months
- The assessment technique
- Continuing positive recognition

Appendix 10: Assessment Questions Used by Supervisors

In the sessions three months ago and today, we talked about a lot of concepts and theories. Use some of these as you look at each individual that works for you:

Which motivator seems to turn him on most?

Achieving?
Responsibility?
Recognition?
Growth?
The work itself?
Advancement?

Which dissatisfier seems to turn him off most?

Company policy?

Rules?
Me?
My approach?
Working conditions?
Money, security?

How important is the peer influence on him?

Is he a part of the group or a loner?
Is it a strong cohesive group, or a weak one?
Is his group for safety, or do they fight it?

How have I been managing him?

Authoritatively?
Participatively?

How should I be managing him if I want to turn him on?
To him, am I credible when I push safety?

Does he think I know what I'm talking about?
Does he think I have a right to talk safety based on the past?

What attitude toward safety does he think I have?
What attitude toward safety does he have?
How do I measure him most of the time?

By the amount of work?
How safe he works?
How fast he works?
Some other way?

How do I reward him? How should I?
What do I reward him for most of the time?
What priority does he think safety is? Does he think I have?
What is his perception of the climate of our organization?

Appendix 11: Analysis of Responses to Pilot Survey Questionnaires for Railroads I and II. Source: Based on American association of railroads

Shown below are composite scores for 18 key areas of safety performance for both railroads. Responses to questions, related to each key area, were grouped to provide the composite scores represented by the length of the line for each railroad. Seven areas in which significant differences in response (10 or more points) occurred are asterisked. These areas may well be the most critical to the success of a safety program.

% Positive

Accident investigation
Accident proneness-method of handling
Alcohol and drug abuse
* Attitudes toward safety
* Communications–safety
Employment practices
* Goals for safety performance
* Hazards corrections
Inspections
* Involvement of employess
Motivational programs
Recognition for good safety perf.
* Rules violations-discipline
Safety meetings
Safety regulations
Supervisor training
* Support for safety-organization
Training

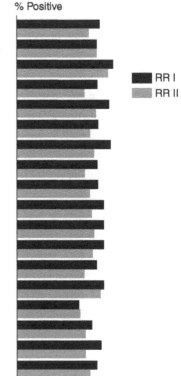

RR I
RR II

The composite score supports hypothesis that the railroad with the highest positive response will demonstrate the best safety performance by other measures as well.

% Positive

Composite score - all areas

ASSOCIATION
OF AMERICAN
RAILROADS

C

The Charter Document

An important outcome of this work is the creation of a **Charter Document** for the steering team, which must be completed by the third steering team meeting. The Charter Doc serves as a guide to the ongoing work of the steering team. Most of the elements completed during the strategic planning process will be included in the final charter document. The final Charter Document should include the following components:

- Steering team POP statements.
- Scope of work.
- Authority.
- Deliverables.
- Measurements – leading and lagging indicators.
- Resources available.
- Executive sponsorship.
- Profile of the steering team.
- Roles and responsibilities.
- Terms of services and rotating membership.
- Schedule and meetings.
- Action item matrix.
- Critical success factors.
- Recognizing success.
- Communication.
- Selection of CI Team topics.
- Selection of CI Team members.
- Onboarding of steering team members.
- Annual review of the Charter Document.

Note: The Charter Document must be approved by the Executive Team.

Delivering Safety Excellence: Engagement Culture at Every Level, First Edition. Michael M. Williamsen.
© 2021 John Wiley & Sons, Inc. Published 2021 by John Wiley & Sons, Inc.

C.1 Process and Objectives (Outcomes)

Assess
- Identify the employee perceptions toward safety (to be measured through a repeated
safety perception survey).

Plan
- Establish a strategic plan (charter) to achieve safety excellence.
- Create and sustain a positive working relationship with the safety department and all other stakeholders.
- Ensure sustainability and continuous improvement by engagement and involvement of our employees.

Develop
- Leading indicator metric system that measures the quantity and quality of defined safety activities.
- Communication plan for the safety culture improvement journey.
- Learning plan for the team and organization.
- Appoint continuous improvement teams as needed.

Implement
- Improve employee perceptions toward safety (to be measured through a repeated safety perception survey (SPS) with a minimum three-year frequency).
- SPS will be completed and results communicated to Safety Steering Team within five months of roll out.
- Engage senior leadership to communicate to the organization the upcoming SPS activity, and a system by which the message will be cascaded through the organization's ranks, ensuring all employees are reached.
- Communication from senior leadership to the organization (and system to cascade message to all employees) thanking employees for their participation in the SPS and providing preliminary information of the results.
- Create and sustain a positive working relationship with the safety department and all other stakeholders.
- Steering team will meet on a monthly basis, following an established agenda.
- Establish at least one CI team per year to error proof existing safety processes.
- One steering team member will be a participant on each CI team.

C.2 Scope and Authority

This steering team charter applies to all employees, contractors (where applicable), processes, and materials within the company. The steering team has the obligation to plan and execute initiatives as deemed necessary, as it applies to the zero risk activity initiative process.

C.3 Roles and Responsibilities

The following are roles of steering team and its members:

- Actively participate
- Attend all meetings
- Carry out all assignments
- Support team decisions
- Share knowledge and expertise
- Help collect and analyze data
- Sponsor CI teams.

C.4 Team Member Representation

The team shall consist of a diverse group of 11 members that provide a functional cross section from facilities and different levels of employees, which will include 2 senior level managers and 1 recorder. Any replacement of team members will be agreed upon by steering team members.

C.5 Team Safety Department Representative

One member of the safety department will attend all meetings as a resource and nonvoting member.

C.6 Voting and Quorum

Eight members must be present to hold a vote. Eight positive votes are required to pass a motion.

C.7 Team Member Service

The steering team members may serve as long as they wish but must be willing to commit for two years.

C.8   Team Leader Service

Team leader will serve for a period of two calendar years, reviewed by team members after two years. Team members must be active for one year prior to becoming the Team Leader. Team Leader will continue for one year as a mentor for the new Team Leader.

C.9   Selection of Team Leader

The team leader must be nominated and voted on by the team.

C.10   Meeting Frequency

The steering team will meet on a monthly basis, following an established agenda.

C.11   Recordkeeping

The team leader is responsible for arranging for the following documentation:

- Facilitate effective Team Meetings.
- Coordinate team activities.
- Ensure adequate attendance to each meeting by coaching team members and removing roadblocks.
- Must be active team member for one year.
- Encourage active participation by each member.
- Help keep team focused by not allowing "scope creep."
- Ensure Actions Items are complete.
- Review and make sure agenda is created and communicated prior to each meeting.
- Responsible for communication to management.

    Standing agenda items are:

- Share a positive safety moment
- Continuous improvement team progress reports
- Facility updates
- Review open action item matrix
- New items
- Open discussion
- Next meeting date and time.

The team leader/recorder is responsible to ensure all team documents are filed within the common folder, according to established recordkeeping protocol. The team defines records as the following (at a minimum):

- Agenda
- Meeting minutes
- Action item matrix
- Parking lot document
- Team charter
- Defined activities of continuous improvement teams
- Presentations
- Perception survey results
- Audit results.

C.12   Communication

The steering team will follow, review, and update the zero risk activity Communication Plan.

C.13   Team Learning Plan

Time will be set aside on the agenda for continuous team learning

C.14   Annual Review of POP Statement (Purpose Objective Process) and Team Charter

The team will annually review (usually midyear) the POP and Charter with the following characteristics of an effective vision:

- Imaginable: Conveys a picture of what the future will look like.
- Desirable: Appeals to the long-term interests of employees, customers, stockholders, and others who have a stake in the enterprise.
- Feasible: Comprises realistic, attainable goals.
- Focused: Is clear enough to provide guidance in decision-making.
- Flexible: Is general enough to allow individual initiative and alternative responses in light of changing conditions.
- Communicable: Is easy to communicate; can be successfully explained within five minutes.

C.15   Measurables

While there are a number of lagging indicators available, the steering team will explore and recommend leading indicators that will tie in to the company and its subsidiaries' strategic plans and tie in to the continuous improvement processes as they are developed.

C.16   Effective Team Norms

- Arrive on time, ready to work, with all assignments completed.
- Start on time and end meeting on time.
- All members participate with commitment.
- Respect all team members.
- One voice, no outside meetings. Every member needs to stand by team decision.
- Manage all distractors.
- Remain focused, stay on task, keep relevant.
- Complaint = goal.
- Have fun, remain positive.
- Open minded, slow to speak quick to listen.
- Be tough on issues not on people.
- Be honest and truthful to team.
- Do what is best for our company.
- Be accountable.
- Team member responsibility to cover role if absent.
- Agreement is majority.

C.17   Steering Team Member Training

A steering team member will meet with all new members prior to their first SST meeting to review the Team Charter and other pertinent information.

C.17.1   CIT Facilitator

Will have two persons trained as a facilitator.
  Facilitator's role

- To facilitate the four day RIW.
- To attend regular CIT meetings.
- To mentor the CIT.
- Liaison between SST and CIT.
- To help identify and mentor a second facilitator.

C.18   Continuous Improvement Team Management

The steering team will select members including a recorder for continuous improvement (CI) teams, in consultation with management, that provide a reasonable representation of the groups affected by the initiative. A minimum of one member of the steering team will sponsor each continuous improvement (CI) team. The steering team sponsor will play a support role for the continuous improvement (CI) team. The steering team will commit to do the following process:

- Hand deliver invitation letter by steering team Member and/or Immediate Supervisor.
- Set up four-day RIW for CI team.
- Give reminder one week prior to RIW.
- Assist in making motel, meals, and other arrangements.
- Create the folder and give access to CI team members prior to first meeting.
- Ensure each CI team has a recorder assigned to them for the RIW.
- All CI teams will include a minimum of one CI team member from a previous team.

C.19   Continuous Improvement Topics

Continuous improvement (CI) team focus topics, or initiatives, will be selected from the Safety Survey results or as identified by the steering team.

C.19.1   Continuous Improvement Process Implementation and Sustainability

Every CI team will make recommendations to the SST for sustainability including a "check" of their implemented process.

SST will follow up on the process to confirm sustainability (feedback, survey, etc.).

SST will develop a Company governing document for each process.

Every CI team project will be piloted before implemented.

Index

Delivering Safety Excellence: Engagement Culture at Every Level, First Edition. Michael M. Williamsen.
© 2021 John Wiley & Sons, Inc. Published 2021 by John Wiley & Sons, Inc.